.

Emanation, Emergence, and Eucatastrophe

Book 7 of Physics from Maximal Information Emanation, a seven-book physics series.

ISBN 979-8-9888160-0-3

Emanation, Emergence, and Eucatastrophe

by

Stephen Winters-Hilt

ISBN 979-8-9888160-0-3

Golden Tao Publishing
Angel Fire, NM
USA

Dedication

This book is dedicated to my family that helped on this lengthy road of discovery: Cindy, Nathaniel, Zachary, Sybil, Eric, Joshua, Teresa, Steffen, Hannah, Anders, Angelo, John and Susan.

Contents

Preface to Physics Series on:

Physics from Maximal Information Emanation

> "The Road goes ever on and on
> Down from the door where it began.
> Now far ahead the Road has gone,
> And I must follow, if I can,
> Pursuing it with eager feet,
> Until it joins some larger way
> Where many paths and errands meet.
> And whither then? I cannot say"

— J.R.R. Tolkien, The Fellowship of the Ring

Variation, Propagation, and Emanation

This is a seven book Physics Series that starts with Classical Mechanics (Book 1 [103]), then Classical Field Theory, such as electromagnetism (Book 2 [104]), then Manifold Dynamics, such a General Relativity (Book 3 [105]). The switch to a quantum mechanics description is given in Book 4 [106], and to a quantum field theory, QED in particular, in Book 5 [107]. A 'quantum manifold theory' would be the obvious next step except it cannot be done (there is not a renormalizable Field theory for Gravitation). Instead a thermal quantum manifold theory is considered, as well as Black Hole thermodynamics in general, in Book 6 [108]. Book 7 [109] describes a new theory, Emanator Theory, that provides a deeper mathematical construct that undergirds Quantum theory, much like quantum theory can be shown to provide a deeper (complexified) mathematical construct based on the classical theory.

This is a modern exposition where subtleties of chaos theory are described in Book 1, of Lorentz Invariance in Book 2, of Covariant Derivatives (General Relativity) and Gauge Covariant Derivatives (Yang-Mills Field Theory) in Book3. Book 4 on Quantum Mechanics provides an extensive review of QM, then considers a full self-adjoint analysis on the full general relativistic solution to the spherical shell in-fall system (a result carried over from Book 3). Book 5 considers QFT basics in detail, along with alternate vacua in specific scenarios. Book 6 considers thermodynamics from the basics to the Hamiltonian thermodynamics of some Black Hole systems. Throughout, the odd recurrence of the alpha parameter is noted. In Book 7 we look to a deeper mathematical formulation from which the Quantum Path Integral formulation would result, as well as explaining the odd parameters and structures that have been discovered (such as alpha and Lorentz Invariance).

The physical description starts with the classic formulations of point particle motion. The first approach to doing this is using differential equations (Newton's 1st and 2nd Law); the second is using a variational function formulation to select the differential equation (Lagrangian variation); the third is using a variational functional formulation (Action formulation) to select the variational function formulation. Historically, it wasn't realized until much later that there are two domains for motion in many systems: non-chaotic; and chaotic.

In a description of particle motion, assuming not in a parameter domain with chaotic motion, several important limits are found to exist. Examples include: the universal constants from the aforementioned chaos phenomenon, that are still encountered in non-chaos regimes if driven "to the edge of chaos". Limits are found where scattering is defined in the asymptotic limit and perturbation theory is well-defined in the sense that it is convergent. Overall, if the evolution is described as a 'process' it is often a Martingale process, which has well-defined limits. So, we have descriptions for motion, typically reducible to an ordinary differential equation (ODE), and for which solutions (requiring limit-definitions) are typically found to exist.

The physical description then contends with field dynamics in 2D, 3D, and 4D (in Book 3 [105]). Two-dimensional ("2D") field dynamics can be described as a complex function (that maps complex numbers to complex numbers). A novelty of the 2D complex function is it also shows how to handle many types of singularities (the residue theorem), thus provides important information about fundamental structures in physics as well as fundamental mathematical techniques for solving many integrals. For the 3D field dynamics we do an analysis of the electromagnetic field in 3D. The level of coverage begins at an overview of electrostatics at the level of the graduate text Jackson [110]. Some problems from Jackson Ch's 1-3 are examined closely in developing the theory itself. For some this material (in Book 2 [104]) might provide a useful accompaniment to Jackson's text in a full course on EM (based from Jackson's text). A quick review of electrodynamics and electromagnetic wave phenomena is then given. In essence, we see many more examples of ODE problems with solutions, such as for the 3D Laplacian, usually involving separation of variables. We then review the famous transform, discovered by Lorentz in 1899 [111], that relates the EM field as seen by two observers differing by a relative velocity. With the existence of this transform, that brings in the time dimension along with the relative velocity, we effectively have a 4D theory.

From Lorentz Invariance we have, as a point transformation, rotational invariance under SO(3) or SU(2). If Lorentz Invariance is fundamental, then we should see both forms of rotation invariance, one of vector/tensor type from SO(3), and one of spinorial type from SU(2). This is the case, as gauge fields are vectorial and matter fields are spinorial. From Lorenz Invariance as a local invariance we have the Minkowski (flat) spacetime metric, which then generalizes to the Riemannian metric (in General Relativity).

As with the point particle dynamics, for the field dynamics we have three ways to formulate the behavior: (1) differential equation; (2) function variation (on Lagrangian); and (3) functional variation (on the Action). We will see similar limit phenomena as before, but also new phenomena, including (i) inevitable BH singularity formation (the Penrose singularity theorem); (ii) FRW Universe formation (from homogeneity and isotropy); (iii) the BH collapse singularity; (iv) the atomic collapse radiative 'singularity'.

Classical dynamics, thus, has two field-like formulations to describe the world: field and manifold. Such formulations can be interrelated mathematically, so what is happening is more a matter of physics emphasis and convenience. The emphasis on this difference, that appears to be no difference (mathematically), is that different physical phenomenologies are at play. Field descriptions appear to work for 'matter', where the fundamental elements are spinorial. Manifold descriptions appear to work best for geometrodynamics (GR), where the fundamental elements are vectorial (or tensorial, such as the metric). Matter fields are renormalizable, thus quantizable in the standard QFT formulation (to be described in Book 5 [107]), while gravitational manifolds are not renormalizable, and have constraints (weak energy condition and positive energy condition given the existence of spinor fields on the manifold).

The presentation in Books 1-3 [103-105], on 'classical' physics, is partly done to make the transition to quantum physics simple, obvious, and in some cases, trivial. Consider the functional variation (Action) formulation of the behavior (whether point-particle or field), this can be captured in integral form, as was done by D'Alembert very early [112] (then by Laplace [113]). Note the use of a large constant to effect a 'highly damped' integral for selection purposes (on variational extremum of the action). To transition to the quantum theory we also have the large constant from 1/h, and so the only difference is the introduction of a factor of 'i', to effect a 'highly oscillatory' integral for selection purposes.

After the transition to a quantum theory, for the point-particle descriptions, the classical collapse problem for atomic nuclei is eliminated. The spectral predictions have excellent agreement with theory, but there is still fine-structure in the spectra not fully explained. The theory is not relativistic and some initial corrections for this are possible (without going to a field-theory) and these indicate closer agreement and explain most of the fine-structure constant discrepancy (and reveal alpha in another place in the theory). It is shown in Book 3 [105] and Book 4 [106], that the GR singularity problem, however, remains unresolved (for the test case of spherical dust shell collapse, done in a full GR analysis, then quantized in a full self-adjoint quantization analysis [106]).

In Book 5 [107], the transition to quantum theory is continued to the field theory descriptions. A precise description/agreement of atomic nuclei is now possible with QED, and within the nuclei themselves (quark confinement) with QCD. The field theories have a small set of bothersome infinities, however, which is eventually solved by renormalization [107]. As mentioned, the quantization of manifold theories, such as GR, does not appear to be possible due to non-renormalizability. Not to be deterred, in Book 6 [108] we consider a Hamiltonian description of a GR system whose quantization would involve an energy spectrum based on that Hamiltonian, if we then use analytic continuation to take us to the thermal ensemble theory based on the partition function that results, we can consider the thermal quantum gravity (TQG) of such systems.

This last example (from Book 6), showing a consistent TQG theory if we use analyticity, is part of a long sequence of successful maneuvers involving analytic continuations in different settings. What is indicated is the presence of an actual complex structure to the stated theory. There is the trivial complex structure extension mentioned above that brought us from the standard classical physics theory to the standard path integral quantum theory. But we also see actual complex structure at the component level with time complexation (that ties to thermal version of the theory by defining the partition function), and we have complex structure as the dimension-level in the form of the successfully applied dimensional regularization procedure used in the renormalization program.

As well as covering the breadth of core physics topics at both undergraduate and graduate level (for courses taken at Caltech and Oxford), including extensive presentation of problems and their solutions, the Series also examines, in specific cases, the boundaries of the physical world "from the inside" (and then later "from the outside"). To this end exploration of spherical dust collapse to form a singularity is examined in a fully general relativistic formalism, and then carried-over to a quantum minisuperspace (quantum gravity) analysis (in Books 3 and 4 [105,106]). Also examined in-depth are the topics of black hole thermodynamics and quantum field theory with alternate vacua (part of Books 5 and 6 [107,108]). The in-depth material comprises the topics covered in my PhD dissertation [114], portions of which are published [72-74,90].

In recent work on machine learning, that includes statistical learning on neuromanifolds [94], we find a possible new source for a foundational element for statistical mechanics (entropy) via seeking a minimal learning process/path on a neuromanifold [91-93]. By the time the Series reaches thermodynamics in Book 6, therefore, the foundational thermodynamics elements have all been established from the physical descriptions discovered in Books 1-5, they just haven't been put together in a comprehensive analysis that gives us the fundamental constructs of thermodynamics and statistical mechanics. That said, it would seem that thermodynamics is, thus, entirely derivative from other, truly fundamental theories. Not so, in the joining of the parts to make thermodynamics we have something greater than the sum of the parts. In the 'system' descriptions we find that emergent phenomena exist. This, at least, is unique to thermodynamics, so it is fundamental in this "sum greater than the parts' aspect.

In Book 7 (the last) of the Series, we consider the standard physical world, described by modern physics, "from the outside." In doing this we've already eliminated part of the mystery of entropy by the geometric 'neuromanifold' description. If we can understand other oddities of the standard theory, and arrive at them naturally, then we might have an even deeper dive into modern physics, testing the limits of what is possible, and see possible future developments and unifications of the theory. This is what is described in papers [1,6,15,16,22,37,46,58,76,115], and organized along with current results into the final Book of the series.

Efforts in the last book of the Series involve choices and concepts identified in the prior six books of the Series, and theoretical maneuvers gleaned from the most advanced courses in physics and mathematical physics taken while at Caltech (as an undergraduate and then as a graduate) and the Oxford Mathematics Institute (as a graduate), and the University of Wisconsin at Milwaukee (as a graduate).

The broad range of topics covered in the Series is, initially, similar to the Landau & Lifshitz graduate textbook series (see [16]), with a similar exposition on classical mechanics at the start of Book 1. Even with well-established classical mechanics, however, there are significant, modern, updates, such as (modern) chaos theory. In the final two books of the Series (Books 6 and 7 [108,109]) we arrive at statistical mechanics and thermodynamics, together with modern topics such as black hole thermodynamics, thermal quantum gravity, and emanator theory.

Key constants and structures of physics, their discovery from the experimental data, and their theoretical placement in the "Grand Scheme," are emphasized throughout the Series. The constant alpha, a.k.a. the fine structure constant, appears in numerous settings so special note of the occurrence of alpha will be made in each chapter. This is the case even at the outset with Book 1, due to fundamental numerical constants appearing from chaos theory. In Book 7 we see the origin of alpha, as a maximal perturbation amount, appears naturally in a formalism for maximal information 'emanation'. But maximal perturbation in what space and in what manner? In Book 7 of the series [109] we will see a possible representation of such an information entity, and its space of existence, in terms of chiral trigintaduonions.

Thus, in the end, this is an effort to tell of a journey to a special place "where many paths and errands meet", giving rise to emanator theory and an answer to the mystery of alpha. Part of this journey is equivalent to 'finding the arkenstone' (alpha) in the most unlikely of places, the trigintaduonion emanation mathematics underpinning the emanator formalism (e.g., Smaug's Lair, described in Book 7 [109]). Why I should have wandered into such an odd place (mathematically speaking), and why I should posit a deeper form of quantum propagation using hypercomplex trigintaduonions, here called emanation, is why there is such extensive background on standard topics. This extensive background even impacts the classical mechanics description via its modern chaos theory material (due to a possible relation between C_∞ and alpha). The critical role of emergent phenomena is only understood at the end, including for manifolds in geometry and neuromanifolds in statistical mechanics, and leads to a Book 6 that goes from very basic (initial thermodynamics) to very advanced (emergent phenomena). Much is made clear with emanator theory, including how reality is both fractal and emergent. At this point in the journey, as with Tolkien, this much I can say: "The Road goes ever on and on … And whither then? I cannot say".

The seven books in the Series are as follows:
    Book 1. Classical Mechanics and Chaos
    Book 2. Classical Field Theory
    Book 3. Classical Manifold Theory
    Book 4. Quantum Mechanics and the Path Integral Foundation
    Book 5. Quantum Field Theory and the Standard Model
    Book 6. Thermal & Statistical Mechanics, and Black Hole Thermodynamics
    Book 7. Maximum Information Emanation and Emanator Theory

Overview of Book 1
Book 1 is a modern exposition of classical mechanics, including chaos theory, and including ties to later theoretical developments as well. The exposition consists, throughout, of the presentation of interesting problems with many solved, the others left for the reader. The problems are drawn from classical mechanics (CM) and mathematics courses taken at Caltech, Oxford, and the University of Wisconsin. The courses range from undergraduate level to advanced graduate level. The courses had a rich and sophisticated selection of textbook and reference material, as you might expect, and those reference texts are, similarly, drawn on here. Those classical mechanics texts, listed by author, include: Landau and Lifshitz [116]; Goldstein [117]; Fetter & Walecka [118]; Percival & Richards [119]; Arnold (ODE) [120]; Arnold (CM) [121]; Woodhouse [122]; and Bender & Orszag [123]. Notice how the first Arnold reference and the Bender and Orszag reference involve textbooks focused on

ordinary differential equations (ODEs). Likewise, an analysis of the excellent, and rapid, exposition by Landau and Lifshitz, reveals that it partly progresses through the material by going through ODEs of increasing complexity (corresponding to more complicated pendulum motion, for example, such as by adding a frictional force). This strong alignment with the underlying mathematics of ODEs is continued in this exposition, so much so that an appendix is provided for a quick review of ODEs from the applied mathematics perspective.

Particle dynamics, with and without forces, are described, with all arriving at descriptions with chaotic motion, with chaos described in the latter half of Book 1 [103]. Universally it is found that systems transitioning to chaotic behavior do so with a remarkable period-doubling process and this will be described both mathematically and with computer results. In the analysis of such dynamical systems we will find that periodic physical systems can be described in terms of repeated "mappings", e.g., classic dynamic mappings [124], and when described in this way the transition to chaos is made much more mathematically evident (as will be shown). The familiar Mandelbrot set is generated by such a repeated mapping, where it's "edge of chaos" is defined by the fractal boundary of the classic Mandelbrot image.

Properties of the classic Mandelbrot set will be relevant to the physics discussed in Book 1 and Book 7, including the property that the fractal boundary has a fractal dimension of 2 (the fractal dimension of the boundary can be between 1 and 2, to get equal to 2 is special). With the Mandelbrot set we also recover the well-studied constants associated with the universal Feigenbaum constants [2]. In the Mandelbrot set we can clearly see the fundamental constant for maximum perturbation that is at maximum antiphase (negative) with magnitude C_∞, where the same results hold for a family of basic formulations (for a variety of Lagrangian formulations, for example).

From the Lagrangian variational formulation of 'action' for particle motion we will eventually define the path integral functional variational formulation involving that same Lagrangian to arrive at a quantum description for the non-relativistic quantum particle motion (described in detail in Book 4 [106], and relativistic in Book 5 [107]). From the quantum description we arrive at the propagator formalism for describing dynamics (this exists in the classical formulation too, but typically is not used much in that context). Complex propagators will then be found to have ties to statistical mechanics and thermodynamics properties (Book 6 [108]). The ties to statistical mechanics are further emphasized when at the "edge of chaos" but with the orbit motion still confined. This may be associated with an ergodic regime, thus an equilibrium and martingale regime, the existence of which can then be used at the start of Book 6 [108] statistical mechanics and thermodynamics derivations with the existence of equilibria established at the outset. The existence of the familiar entropy measures are already indicated in the neuromanifold description (Book 3 [105]), thus, together with equilibria, the Book 6 thermodynamics description is able to begin with a well-established foundation that is not claimed by fiat, rather claimed as a direct result of what has already been determined in the theory/experiment described in the previous books of the Series.

Overview of Books 2 & 3
When moving from a theory of point particles to a theory of fields, there's not much discussion in the core physics books on fields in a general sense, it usually just directly jumps to the main field of relevance, Electromagnetism (EM). If advanced, it may also cover General Relativity (GR), as with [125]. In what follows we will cover these topics, but we will also cover the more basic fields in 1, 2, and 3D (including fluid dynamics), as well as 4D Lorentzian Field formulations (for Special Relativity), the Gauge Field formulation (thus Yang Mills covered in a classical context), and the GR geometric and gauge formulations. This establishes the foundation for the standard forces, and upon quantization (Books 4 and 5 in the Series), lays the foundation for the standard renormalizable forces (all but gravitation).

The gravitational coupling constant 'G' is a dimensionful coupling (not like with alpha in EM), and gravitation with manifold construct can be described as a gauge field construct, although not renormalizable. Gravitation, and associated geometry/manifolds, appears to relate to its own emergent structure, as will be discussed in Book 6. From the local Lorentzian geometry and Lorentzian field descriptions we also see the first of many examples where there is system information in the complexification of some parameter, here the time component. If the Lorentzian is shifted to complex time, this shifts it to being a Euclidean field, with formally well-defined convergence properties (as occurs in statistical mechanics). Complex time also shows deep connections between classical motion and associated Brownian motion (where random walk reveals pi). Thus, it should not be surprising that an emergent manifold may have complex structure such that there is also an emergent 'thermal' manifold, possibly the neuromanifold described in Book 3 and the related partition functions examined in Book 6. Just like locally flat space-time is a natural construct in GR, so too are optimization "learning" steps on a neuromanifold such that relative entropy is selected as a preferred measure, and from it Shannon entropy and Boltzmann's statistical entropy. Thus, the manifold construct appearing at Book 3 has far reaching impact into the foundations of the thermodynamic and statistical mechanical theory described in Book 6.

Before we even get to the manifold/geometry complexities of GR, however, we have already established much with the EM field part of the theory: (i) from 'free' EM without matter we get the speed of light c, Lorentz invariance, and from that special relativity and locally flat space-time; (ii) from EM with matter we get the dimensionless coupling constant alpha.

In going over field theories to describe matter, force fields, and radiation we first describe the classical field theories (CFTs) of fluid mechanics, EM, and General Relativity, with many examples shown. This is then carried over to the quantum field theory (QFT) description in Book 5. A review of the core mathematical constructs employed in CFT and QFT is given in the Appendix. Even as the mathematical physics approach grows in sophistication, we still obtain solutions via variational extrema. Thus, determining the evolution of the system from its variational optimum now becomes the focus of the effort. System 'propagation' from one time to a later time can be described by a propagator. Although a 'propagator' formulation is possible mathematically in classical mechanics (CM) and classical field theory (CF), which are shown, this is usually not done, in favor of simpler representations for the experimental application at hand. As we move to descriptions in the quantum realm, however, the use of the propagator formalism becomes typical, and when used in the path integral formulations we arrive at a compact formulation describing both the evolution and stationary-phase solution at once.

In Book 2 the focus is on classical field theory in a fixed geometry, the main physical example is EM. In this setting alpha appears, for example, in the description of an electron-positron pair: $F = e^2/(4\pi\varepsilon a^2)$ for electron-positron distance 'a' apart, where alpha appears as the coupling constant. Later, in quantum mechanics (QM), both modern and in the early Bohr model, we have that alpha $= [e^2/(4\pi\varepsilon)]/(c\hbar)$. The appearance of alpha in these situations is occurring in bound systems. If we examine EM interactions that are unbound, on the other hand, such as with the Lorentz Force $F = q(E \times v)$, here there arises no alpha parameter, nor with the early quantum mechanical analysis of such systems such as with Compton scattering. Thus, we see an early role for alpha, but only in bound systems, thus only in systems with (convergent) perturbative expansions in system variables.

In Book 3, classical field theory with *dynamic* geometry, i.e. GR, we don't see alpha at all. Instead we see manifold constructs and the mathematics of differential geometry (and to some extent differential topology and algebraic topology). Manifold constructs are entirely encapsulated in the math background given in Book 3 and the Appendix there. An application in the area of neuromanifolds (see [94]), shows the equivalent of a geodesic path in this setting is evolution involving minimum relative entropy steps. Similar to the description of a locally flat space-time we now have a description of 'entropy' increasing/evolving according to minimum relative entropy.

General relativity (GR) stands apart from the other force fields. All the other force fields are part of an adjoint representation of the standard model vis-à-vis the stability subgroup U(1)xSU(2)$_L$xSU(3). The form of which is derivable from the chiral T one-sided products described in Book 7. The standard model is uniquely obtained in this process, and with no mention of GR. Keep in mind, however, that the adjoint representation has operation on some space (hyperspinorial in case of simple octonion right-products, for example). The 'force' due to gravity is that due to manifold curvature, where the manifold construct is possibly emergent on the space of operation. Thus, the origin of the GR force is entirely different, and it will not allow quantization like the other forces, nor will its singular solutions be resolvable via quantum physics alone, as with EM in Books 4&5, but will also need thermal physics (as will be described in Book 6).

The existence of singular GR solutions, outside of specially symmetric cases (the classic Black hole solutions), wasn't firmly established until the Penrose singularity theorem [126] (awarded Nobel prize in Physics for this in 2020). Some of this material is covered in Book 3 to show how the mathematical formalism shifts to differential topology methods to describe the singularities, with examples referencing the Hawking and Ellis classic [127] and using Penrose diagrams. This, in turn, will come in handy when describing the classic FRW cosmologies with radiation and matter dominated phases (using notes from Peebles [128], Peebles won the Nobel in Physics in 2019).

The GR development would be remiss if it didn't briefly delve into cosmological models, the classic FRW cosmologies in particular. With the GR tools developed, cosmological results are examined, starting with the entry of the cosmological constant into the formalism (a candidate for Dark energy). Various observational data on galaxy rotations and universe simulations of galaxy cluster formation both indicate the existence of Dark matter. This, then, means we have new matter, non-interacting except gravitationally, and this is actually consistent with the latest observational data on the muon g-2 value [129], where the discrepancy between theory and experiment has grown to 4.2 standard deviations, where an extension in the Standard Model appears to be in the works. This is convenient as Emanator theory (Book 7 [109]), predicts such an extension.

We can thus arrive at field equations for EM, GR, and Yang-Mills Gauge Fields (Strong and weak). We can obtain wave and vortex phenomena (as hinted in fluid dynamics). We show the classical instability for atomic matter (classical EM instability) and classical gravitational instability (leading to black hole formation with singularity). From Lagrangian formulations we can then arrive at a QFT formulation (Book 5). The QFT formulation completes the QM (Book 4) cure of "non-relativistic atomic instability" with the cure of the fully relativistic atomic description of the radiative-collapse instability. Introduction of QFT also leads to new instability or infinities, but these can be eliminated by renormalization for the EM and electroweak formulations, and the Yang-Mills strong formulation, but not the GR (gauge) formulation. The current theoretical formulation in modern physics has one glaring gap, therefore: a quantum theory of gravitation. Perhaps this is not a missing element, however, if geometry/GR is a derivative phenomenon, like the field of statistical mechanics and thermodynamics appeared as derivative phenomenon when the complexified quantum propagator gives rise to a real (quantum) partition function. The hint of a deeper emanator theory suggests emergent structures of geometry and thermodynamics are arrived at in the process of emanation, with the information emanated being that of the renormalizable quantum matter fields. In Book 7 [109] a precise mathematical meaning will be found for describing maximal information emanation.

Overview of Book 4

By 1834, with Hamilton's Principle, there was a strong foundation for what is now called classical mechanics. By 1905, with Einstein's publication on the photoelectric effect [130], the rules of classical mechanics were being superseded by the new rules of quantum mechanics. The earliest appearance of quantum mechanics, however, began with the various observations of quantization of light, starting with the strange occurrence of spectral lines for hydrogen. The hydrogen spectrum was made even stranger by a precise fit to a succinct empirical formula by Balmer in 1885 [131]. This is the beginning of an amazing period of discovery. The developments of QM from introductory to advanced roughly follows that history.

The early phase of discovery for quantum mechanics moved into the modern quantum mechanics formalism with the discovery of Heisenberg of the successful application of matrix mechanics and the resultant uncertainty principle (1925) [132]. In 1926, Schrodinger showed that the problem of finding a diagonal Hamiltonian matrix in the Heisenberg's mechanics is equivalent to finding wavefunction solutions to his wave equation [133]. An interpretation of the wavefunction was then clarified in 1927 by Born [347]. Dirac developed a manifestly relativistic formalism for the wavefunction and wave-equation for fermionic matter (1928) [135]. An axiomatic reformulation of quantum mechanics was then given by Dirac (1930) [136], laying the foundation for much of modern quantum notation and for critical issues such as self-adjointness. Dirac then described a formulation of a quantum propagation path, with quantum propagator having the familiar phase factor involving the action, in his paper "The Lagrangian in Quantum Mechanics" in 1933 [137]. In essence, Dirac had obtained a single path, in what would eventually be generalized by Feynman to all paths with the invention of the path integral formalism (1942 & 1948) [52,138]. The equivalence of a quantum mechanical formulation in terms of path integrals and the Schrodinger formalism was shown by Feynman in 1948 [52].

In a path integral description, the quantum mixture state, semiclassical physics, and classical trajectories are all given by the stationary phase dominated component. A stationary phase solution that is dominated by a single path is typical for a classical system. Thus, variational methods are fundamental to analysis of physical systems, whether it be in the form of

Lagrangian and Hamiltonian analysis, or in various equivalent integral formulations.

Feynman's discovery of the path integral formalism wasn't solely based on the prior work of Dirac (1933) [137], although by appending that paper to his PhD thesis (1946) its importance was clearly emphasized. Feynman also benefited from work going as far back as Laplace [113] for selection process based on highly oscillatory integral constructions that self-select for their stationary phase component. This branch of mathematics eventually became associated with Laplace's method of steepest descents, then to the work of Stokes and Lord Kelvin, then to the work of Erdelyi (1953) [49-51].

Feynman and others then invented quantum field theory for electromagnetism (QED) during 1946-1949 (more on this later). Extension to electroweak occurred in 1959, and to QCD in 1973, and to the "Standard Model" in 1973-1975. Thus, the impact of the path integral revolution in quantum physics was felt well into the 1970's, but this was only the beginning. At their inception path integrals were examined by Norbert Wiener, with the introduction of the Wiener Integral, for solving problems in statistical mechanics in diffusion and Brownian motion. In the 1970's this led to what is now known as "the grand synthesis" which unified quantum field theory (QFT) and statistical field theory (SFT) of a fluctuating field near a second-order phase transition, and where use of renormalization group methods enabled significant advances from QFT to be carried over to SFT.

The grand synthesis is one of many instances to come where we see analytic continuation of a constant or a parameter giving rise to familiar physics in the thermodynamic and statistical mechanics domains, showing a deeper connection (still not fully understood, see Book 7). The Schrödinger equation, for example, can be seen to be a diffusion equation with an imaginary diffusion constant [139,140]. Likewise, the path integral can be seen to be an analytic continuation of the method for summing up all possible random walks [141].

In Book 4 we also carefully examine the closest gravitational equivalent to the hydrogenic atom (dust shell collapse). What results is an incomplete formulation due to boundary conditions, where to get the time choice you must input that time choice. No specific choice of time is indicated to avoid infall-collapse. The results, however, can show stability and consistency in a "full" thermal quantum gravity description where analyticity is employed. Success in this way, and not others, suggests possible fundamental role of analyticity and thermality (Books 6&7) and also suggests that thermal quantum gravity TQG may 'exist' or be well-formulate-able, while quantum gravity QG generally might not 'exist'. These results, shown in Book 6, provide the lead-in to the Book 7 discussion on Emanator theory, where core concepts in Books 1-6 that tie to emanator theory are brought together in a new theoretical synthesis.

Overview of Book 5
In Book 5 we show QFT's in the gauge field representation, which clearly relates the choice of field theory to a choice of Lie algebra, which, in turn, can be related to a choice of group theory (such as U(1) and SU(3)). From this we can see that non-classical algebraic constructs are ubiquitous in QM and QFT, so a review of Group Theory and Lie Algebras is given in the Appendix, as well as a review of Grassman Algebras, and other special algebras needed in QM and QFT. Similarly, as regards choice of approach, we find that the Schrodinger and Heisenberg formulations often provide the only tractable way to get a solution for bound systems. In critical theoretical considerations, however, the path integral approach is best (as will be shown). In seeking a deeper theory, the more unified path integral (PI) approach provides important hints as to a deeper theory (see Book 7).

In Book 5 we get the highest precision result for the value of alpha, in its role as perturbation parameter. If a calculation of the electron magnetic moment parameter g-2 is performed, with all of the Feynman diagrams appropriate to expansions up to 5th order, we get a determination of alpha up to 14 digits, where 1/alpha=137.05999..... . This gives us one of the most precise measurements of alpha known. When a similar analysis is done for the muon g-2, given the much larger muon mass, particle production pairs of other particles have a measurable effect, and we are able to probe the lower masses of the standard model that are present. In doing this, in preliminary experiments, there is a discrepancy indicating more particles, e.g. the Standard Model will need to be extended (possibly with a type of 'sterile' neutrino). These missing particles could be the missing "Dark Matter". The prediction of such in Emanator Theory, and why there should be an imbalance between the left and right neutrinos (hint: maximum information transmission) is described in Book 7.

Part of the description of quantum field theory entails use of analyticity and other complex structures to encapsulate more of the physics in a complex extension to the space (or dimension). This often leads to formulations in terms of complex integration, with the choice of complex contour specified, such as with the Feynman propagator. One of the main renormalization methods, for example, is to use dimensional regularization, which entails analytically continuing expressions with dimensionality to dimensionality as a complex parameter. There is also the aforementioned shift to complex and to "Wick rotate" expressions with real time to expressions with pure complex time. In doing this the statistical mechanical partition function for the system is obtained, with well-defined summation. Thus, a connection between 'thermality' and complex structure, in the time dimension at least, is indicated.

The second part of Book 5 describes QFT on curved space-time (CST), where we arrive at an early analysis of Black Hole thermodynamics. Here we find that space-time curvature gives rise to thermality and particle production effects. Black Hole thermality was revealed in Hawking radiation [142], due to the causal boundary at the horizon. Such thermality is even seen in flat space-time (Book 5) if causal boundaries are induced, such as in the case of an accelerated observer [143].

QFT on CST has one further gift, critical to the statistical mechanics formalism to follow in Book 6, and that's the spin-statistics relation. This relation is usually assumed, along with other critical notions, such as entropy, and the relation between entropy and density of states. These are all shown, with the presentation path chosen in this Physics Series, to be fundamental or derivative to the formalism already established in Books 1-5 (to prepare for Book 6).

The choice of time is related to choice of vacuum, which is related to choice of field geometry or observer motion (such as constant acceleration or expansion). If you have flat spacetime QFT with a boundary, then you have thermodynamic effects (e.g., the Rindler observer). In this setting we can compare the Hawking derivation of Hawking Radiation using the Euclideanization 'trick' vs the Bogoliubov transformations of the field to the Rindler geometry from the Minkowski geometry (if chosen as the asymptotic vacuum reference). With QFT on CST we also arrive at spin-statistics as mentioned, and get the final extension of the theory by way of Grassman algebras, to arrive at thermodynamically consistent Bose and Fermi statistical descriptions on quantum matter.

Overview of Book 6

Thermodynamics is the oldest of the physics disciplines (fire), with unapologetic use of phenomenological arguments and mysterious thermodynamic potentials (entropy). Obviously, thermodynamics is still prevalent today, including in its more quantified form via statistical mechanics. How is this not a failure of the mechanistic description of the universe indicated by CM and even QM? Concepts that appeared in QM, such as probability, are now occurring again. Other new concepts appear as well, including: approximate statistical laws; equations of state; heat as a form of energy; entropy as a variable of state; existence of equilibria; ensembles/distributions; and existence of the partition function. Many of these concepts appear in the path integral descriptions with the analyticity methods/extensions mentioned previously, so there are hints of a deeper theory that arrives at much of thermodynamics/Statistical mechanics foundation from the existing quantum theory.

Book 6 has been placed after the other chapters to await identification of entropy as fundamental in that it can be identified as an intrinsic system function even before getting to thermodynamics. We also already have experience with many particle systems, via QFT (especially in CST where particle creation is almost unavoidable), without directly tackling that scenario (due to QFT effectively already being many-particle, with analytic determination of many-particle system functions, such as entropy). With entropy presented at the outset as an important system variable, the derivation of thermodynamic potentials is then a straightforward process, as will be shown. The standard SM connections to thermodynamics can then be given. Thus, in covering Thermodynamics and Statistical Mechanics we start with the foundations of the theory mostly established, such as entropy (also with equipartition equivalent to sum on paths with no weightings, etc.), with no assumptions. Everything follows directly from the theoretical discoveries outlined in the preceding books in the Series. We don't see new connections to alpha, but we do see new structures/effects, especially manifold constructs (as with GR, where we also saw no role for alpha).

The close ties between QM Complexified giving rise to a particle ensemble partition function, and QFT complexified and field ensemble partition function, is now simply a derivative aspect of the fundamental complexation posited. This complexation will be posed in Book 7 with emanation in a complexified perturbation space.

From Atomic Physics, described in Book 4, we also obtain the standard rules on electron shell completion (that is encoded in the periodic table). Similarly, we can also understand the origins of the intermolecular quantum chemistry rules. When taken to the statistical mechanics (SM) extreme we have thermodynamic equilibrium emergent from (the Law of Large Numbers (LLN) and reverse Martingale convergence. With completion of application to chemical processes we have clear phase-transition effects, as well as equilibrium and near-equilibrium effects. The familiar chemistry results, with phases of matter.

From chemical equilibrium and near-equilibrium, with 10^{23} elements that interact weakly or not at all, we have two generalizations. The first is to consider chemical near-equilibrium and directly obtain an emergent process at this level, this is the branch that gives us biology/life at its most primitive level. The second is to consider equilibrium and near-equilibrium in general when the elements interact strongly (with 10^{10} elements, say), this is the branch that describes biology/life at its most advanced social level and economics. In classic shot noise, the granularity of low-current flow (due to discreteness off electron charge) leads to a noise effect. Thus, as we consider situations with fewer elements, there are more complications, not less, due to granularity noise effects, and we enter the realm of machine learning with sparse data. Noise effects can be significant in complex systems, especially in biology where it is part of what is selected (such as in hearing, for background noise cancellation).

The second part of Book 6 explores the role of thermodynamics in efforts to extend to TQFT and TQG. This is done by exploring Black Hole settings. The recognition of a role for complex structure on system variables becomes apparent in this process (on top of the generalization to non-trivial algebras as already revealed).

In Book 6, part 2, we examine the Hamiltonian thermodynamics of some black hole geometries with stabilizing boundary conditions. In this foray into directly exploring a thermal quantum gravity (TQG) solution we assume a path integral form for the GR problem and shift directly to a partition function (by 'Wick rotation' mentioned above). We see that TQG is possible, where positive heat capacity shows stability. Another encouraging result as to an eventual unifying theory comes from String theory via its explanation of BH thermodynamics and BH horizon effects with the BH fuzz solution (via use of the holographic hypothesis and the related AdS-CFT relation [100,144]).

In Book 6, part 2, we also examine the propagator to partition-function transformation upon complexation, which leads to a thermodynamic theory for some equilibrium formulation, with certain parameter settings required for stability (positive heat capacity). This is doable in a variety of settings, suggesting how such thermodynamically consistent boundary conditions may be what constrains the classical motion and BH singularity formulation by the effect of this stabilization manifesting for certain internal geometries. Successful TQG (Thermal Quantum Gravity) formulations, such as for RNadS and Lovelock spacetimes shown in Book 6, via reformulation using analyticity, and not via non-analytic approaches, suggests a possible fundamental role of analyticity once again and also suggest that TQG may 'exist' or be well-formulate-able, while QG generally might not 'exist'. These results, together with core concepts from Books 1-6 that tie to emanator theory, are brought together in a new theoretical synthesis in Book 7.

Overview of Book 7
In Books 4,5, and 6 of the Series, we explored examples of QM with imaginary time, QFT in CST, Thermal QFT, minisuperspace QG, and Thermal QG. In this effort we find the path integral, and PI propagator, to provide the most general representation. In seeking a deeper theory in Book 7 we build on the sum-on-paths with propagator formulation to arrive at a sum-on-emanations with emanator formulation.

Propagation in a complex Hilbert space, in a standard QM or QFT formulation, requires the propagator function to be a complex number (not real or quaternionic, etc., [19]). This prohibits what would otherwise be an obvious generalization to hypercomplex algebras. In order to achieve this generalization, we have to introduce a new layer to the theory, one with universal emanation involving hypercomplex algebras (trigintaduonions) that is hypothesized to project to the familiar complex Hilbert space propagation with associated fixed elements (e.g., the emanator formalism projects out the observed constants and group structure of the standard model). The 'projection' is an induced mathematical construct, like having SU(3) on products of octonions, but here it we be the standard model U(1)xSU(2)xSU(3) on products of emanator trigintaduonions. Thus, in Book 7 a unified variational formulation is posed, one that arrives at alpha as a natural structural element, among other things, uniquely specified by the condition of maximal information emanation.

In Book 7 we also make note of the implications of a fundamental mathematical operation on a space that is repeated or added. The non-GR forces are given by the form of the operation (the sequence forming an associative algebra), the GR forces are given indirectly by the form of the space, this leaves the aspect "repeated or added" to be considered with care. If a purely 'repeated' operation, or mapping, occurs we can return to the dynamical mapping discussion of Book 1, where chaos can occur and is ubiquitous. There, the primal 'phase transition', the transition to chaos, is evident. If an operation with addition is involved (in the statistical sense of multiple elements), along with repeated overall steps, we arrive at the general framework of statistical mechanics with effects from the Law of Large Numbers (LLN) and reverse Martingale convergence, among other things (Book 6). Most notable, however, is the prevalence of a new effect, that of phase transitions and the emergence of new structure (order from disorder), including the remarkable structures of chemistry and biology.

Why the recurring 'Cabbalistic formula'? was a question even in the time of Sommerfeld [3]. Now, the numerological parallel is more exact than realized at that time, so is too much a coincidence to be by chance. The non-coincidence appears to be due to the maximal nature of information transmission in a variety of circumstances (in physics, biology, and even human communication with sufficient optimization) as well as with the fractal-like repetition of key parameter sets that occurs in these different settings $\{10,22,78,137 \cong 1/alpha\}$. We see that 10 expresses the dimensionality of propagation (or nodes of connectivity), while 22 corresponds to the number of fixed parameters in the propagation (in Book 7 we explore propagation in a 10 dimensional subspace of the 32 dimensional trigintaduonion space, leaving 22 dimensions at fixed values that appear as parameters in the theory). We will see the number 78 relates to generators of the motion, and that there are 4 chiralities of motion ('doubly chiral'). We will also see that 137 is simply the number of independent tri-octonionic product terms in the general chiral trigintaduonion 'emanation'.

Synopsis – Frodo Lives

Tolkien wrote of eucatastrophes [102], perhaps he anticipated the constructive role of emergent phenomena in maximum information transmission.

Preface to Physics Series, Book #7, on:

Emanation, Emergence, and Eucatastrophe

Propagation in a complex Hilbert space, in a standard Quantum Mechanics or Quantum Field Theory formulation, requires the propagator function to be a complex number. This prohibits what would otherwise be an obvious generalization to hypercomplex algebras. In order to achieve this generalization, we have to introduce a new layer to the theory, one with universal emanation involving hypercomplex algebras (trigintaduonions) that is hypothesized to project to the familiar complex Hilbert space propagation with associated fixed elements (e.g., the emanator formalism projects out the observed constants and group structure of the standard model). The 'projection' is an induced mathematical construct, like having SU(3) on products of octonions, but here it we be the standard model U(1)xSU(2)xSU(3) on products of emanator trigintaduonions. Thus, in Book 7, last of the Physics Series, a unified variational formulation is posed, one that arrives at alpha as a natural structural element, and the standard model, among other things, uniquely specified by the condition of maximal information emanation.

Chapter 1. Introduction

This book is the culmination of the seven-book Series: "Physics from Maximal Information Emanation." The first six books cover the key structures of physics in a modern exposition, often entailing advanced mathematical methods (for which background is provided in the Appendices of the books of the Series). In the first six books fundamental physical/mathematical elements are identified, particularly recurring aspects like alpha perturbation and the manifold construct. This, then, forms the basis for a clear start to the Emanator formalism presented here (much like books 1-5 lay the foundation for many thermodynamic concepts so that upon formally describing thermodynamics in Book 6, much of the foundational elements had already been established).

Emanator theory stems from the Maximum Information Emanation (MIE) Hypothesis. The definition of information is context dependent, so how the MIE hypothesis will manifest depends on circumstance. We start with the fundamental notion of the quantum propagator, for which mathematical 'propagators' satisfy unitarity. We seek to extend this foundational element so start by asking what is the highest Cayley algebra that can remain unitary – where the answer comes down to the highest order division algebra, which are the octonions. What if we change the desired property from unitary to unit-norm preserving? Then we can extend 2 more dimensions beyond the 8D octonion algebra to a chiral subspace of the trigintaduonions that is 10D [1].

The 10D chiral trigintaduonions are then identified as the maximal information 'carriers' or emanators, operationally like the quantum propagators in a larger theory, where evolution of the system will be shown to result from sums on paths of emanators (similar to quantum evolution in terms of a path integral on propagators 'steps'). Having posited this maximal construct we see the classic signs of "asking the right question" since we get a variety of clear results:

(1) the chiral emanator is manifestly Lorentz Invariant.

(2) chiral emanation involves a 10D element in a 32D space (trigintaduonions) for which maximum perturbation (still permitting unit-

norm transmission) is by the amount 'alpha' – we therefore have the mysterious alpha by a computational definition.

(3) chiral emanation with perturbation does not have effective dimension 32D due to chiral and other constraints -- noise budget analysis or (equivalently) Kato Rellich operator analysis, both indicate and effective dimension slightly greater than 29 referred to as "29*".

(4) the chiral emanator indicates 'motion' in a 10D subspace of 32D, suggesting 22 constants of the motion. This is revealed to be true in the mathematical formalism as there are 22 types of emanation that result in no change to the base trigintaduonion describing the system. This, in turn, suggest that the emanator theory will have 22 parameters.

(5) By using the split form of the trigintaduonions, we not only have manifest Lorentz invariance, we also have an exact algebraic split to a space that is simply the direct product of 29* real dimensions (not a local approximation to such). This is important because the fundamental existence of a complex structure (two such layers of structure will be indicated) means that we trivially have the extension $\mathbb{R}^{29^*} \to \mathbb{C}^{29^*}$. Suppose we have point-like singular elements in $\mathbb{C}^{29^*}$, such will occur due to zero-divisors in the 32D trigintaduonion space. To achieve a maximal domain of analyticity (an application of MIE), we must remove the zd-singularities. In doing so we obtain point-like matter in the theory and a small-h constant that enters the sum on emanator paths just as Planck's constant in the sum on propagator paths in the quantum formulation – suggesting that these small-h numbers are related.

(6) There derivations of alpha are obtained: (i) $\{\alpha\}$ based on the maximal perturbation for which chiral emanation retains the unit-norm property; (ii) $\{\alpha, \pi\}$ based on the maximal noise transmission on a chiral emanation path; and (iii) $\{\alpha, \pi, C_\infty\}$ based on the maximal noise transmission on an achiral emanation path (where maximal emanation is at "the edge of chaos" which is defined according to Feigenbaum Universality [2]).

(7) At component level in the emanation product, using 100's of millions of computational steps, we see an excellent asymptotic fit to random walk behavior. Since a random walk is a Martingale process, this strongly suggests that the achiral emanation process is Martingale. In turn, the projected quantum process (standard theory) would retain the imprint of that Martingale process.

(8) The achiral emanator, a sum of achiral emanation paths, can be shown to have the mathematical form $\sum \exp(i\mathbb{H} \times \mathbb{O}) \to \mathbb{C} \times \mathbb{H} \times \mathbb{O}$, which can

be shown to give the gauge theory of the standard model: $U(1) \times SU(2)_L \times SU(3)$. Thus, there is no grand unified theory in terms of gauges that is fundamental (although a GUT may approximately occur at early and late times cosmologically). The 'ugly' product gauge that is observed is precisely what is predicted by emanator theory.

(9) Universal thermality is indicated by application of the MIE hypothesis to the choice of whether 'effective' achiral emanation is associative or not. Obviously a restriction to associativity would lose some information transmission capability, but it would then enable the existence of two added layers of complex structure. If we take the MIE in this context to realize the maximum information transmission overall, then the prediction is for effective emanation that is associative (a Clifford subalgebra at this level) but with the two layers of complex structure added. One layer of complex structure is associated with thermal behavior, one layer of complex structure is associated with quantum behavior (already exhibited).

(10) In the emanation process there is a clear separation between spinorial elements and manifold elements. Manifold elements include geometry and thermality, have no alpha-perturbation effects, and appear to be part of the 'apparatus' from the perspective of the quantum theory.

1.1 The mystery of alpha

The fine-structure constant, α, has been a mystery confounding physicists for over a century. In early work on spectral analysis where it first appeared, Sommerfeld noted the almost cabbalistic underpinnings of the mathematics (in his book Atombau and Spektrallinien [3], Sommerfeld referred to the Rydberg top square equation as a 'cabbalistic' formula). Wolfgang Pauli, a student of Sommerfeld's, shared his keen interest in the origins of α and turned it into a life-long obsession. So much so, that it practically drove him mad, to where he sought the help of famed psychoanalyst Carl Jung, with whom he eventually partnered to try to solve the mystery of α (the madness is contagious). From Pauli's Nobel Prize Lecture [4]:

> "From the view of logic my report on 'Exclusion principle and quantum mechanics' has no conclusion. I believe it will only be possible to write the conclusion if *a theory will be established which will determine the value of the fine structure constant* and will thus explain the atomistic of electric fields actually occurring in nature." (emphasis mine)

3

The obsession with α continued with the next generation of great Physicists as well, particularly Feynman, who said [5]:

"There is a most profound and beautiful question associated with the observed coupling constant, e – the amplitude for a real electron to emit or absorb a real photon. It is a simple number that has been experimentally determined to be close to 0.08542455. (My physicist friends won't recognize this number, because they like to remember it as the inverse of its square: about 137.03597 with about an uncertainty of about 2 in the last decimal place. It has been a mystery ever since it was discovered more than fifty years ago, and all good theoretical physicists put this number up on their wall and worry about it.) Immediately you would like to know where this number for a coupling comes from: is it related to pi or perhaps to the base of natural logarithms? Nobody knows. It's one of the greatest damn mysteries of physics: a magic number that comes to us with no understanding by man. You might say the "hand of God" wrote that number, and "we don't know how He pushed his pencil." We know what kind of a dance to do experimentally to measure this number very accurately, but we don't know what kind of dance to do on the computer to make this number come out, without putting it in secretly!"

1.2 Fractal Reality

Consider maximum "unit-norm" propagation (via right multiplications), e.g., a projection (or 'emanation'), where a hypercomplex 'emanator' has maximum propagation dimensionality ten (a doubly-chiral 10dim subspace of the 32dim space of trigintaduonions). The maximum propagation perturbation allowed from the 10dim space into the embedded 32 dim space is given by the maximum perturbation α for the non-10dim part, where this is taken as the definition of α. Computational efforts to determine the maximal perturbation (while remaining "unit norm") recover the known α [6].

Exploration to high precision indicates a possible fractal limit (as noted in [6]), with possible pattern recurrences as in the Mandelbrot Set on complex numbers. A further complication is that the 32 dim hypercomplex trigintaduonion numbers have also become non-associative (but still retain octonionic sub-space 'braid' rules, which are critical in what follows).

To see the fractal connection, consider the iterative mapping based on the function $z_n = (z_{n-1})^2 + c$. For choice c and initial $z_0=0$, if $z_\infty \to \infty$, then that c is outside the set, otherwise, if remains bounded, then it's in the (Mandelbrot) set. This is an example famous for its beautiful fractal images and mathematical properties. The largest c value (at the edge of chaos) is known as the bifurcation parameter and is $c_\infty = 1.401155189....$

The maximum allowable 'perturbation' for z (not z^2) would then be $(c_\infty)^{(1/2)}$. In the trigintaduonion propagation we discover in what follows we have chiral propagation in the 32 dim trigintaduonion space, where the real dimension is fixed by the unit-norm property, leaving 29 'free' imaginary dimension/parameters, since two more are selected for a specific chirality. If we allow the same maximal bifurcation parameter as a factor for each of the 29 free dimensions (and for an imaginary part overall), a precise relation will be obtained according to the exact form of the trigintaduonion emanation.

In what follows there are, thus, three relations: (i) A computational limit relation $\{\alpha\}$ for α alone due to the fractal limit on chiral trigintaduonions with maximum perturbation; (ii) a relation $\{\alpha, \pi\}$ due to maximum perturbation occurring when noise has maximum antiphase; (iii) the $\{\alpha, \pi, c_\infty\}$ relation due to maximum information flow occurring at the edge of chaos.

1.3 The Maximum Information Emanation (MIE) Hypothesis
The number system, or algebra, used to describe a physical system is typically the real numbers, sometimes the complex numbers (to describe wavelike phase information), and, rarely, the quaternionic numbers (to describe rotation and EM interactions). In recent theoretical efforts, attention has also been paid to octonionic numbers to describe Quantum Electrodynamics (QED) and Quantum chromodynamics (QCD) interactions [8-14]. The algebras given by real, complex, quaternionic, octonionic, sedenionic, trigintaduonionic, ..., are known as the Cayley-Graves algebras, whose dimensions double at each step, one dimension for real, two for complex, four for quaternionic, etc. Maximal *unitary* propagation occurs with the octonion algebra and no higher (thus 'maximal' propagation, seemingly, only in 8 dimensions). What is actually needed in physics 'propagation' is right multiplication with a unit-norm 'propagator', for example, giving rise to a unit-norm result (then iterating to create a path from the infinitesimal propagator steps). If this is sought instead, then a chiral extension can be made from the octonions into the sedenions, and then again into the trigintaduonions, giving rise to a *maximal 'propagation', or projective emanation, in 10 dimensions* within the 32 dimensional trigintaduonions (as shown in [1,6,15,16]).

For Real numbers unit norm propagation is trivial, consisting of multiplying by +1 or -1. For Complex numbers unit norm propagation

5

involves multiplication by complex numbers on the classic unit circle in the complex plane, which reduces to simple phase addition according to rotations about the center of that circle (motions on S^1). For quaternion numbers unit norm propagation is still straightforward since it's still, in the end, a normed division algebra, where $N(xy)=N(x)N(y)$. For the quaternions, instead of motion on S^1, we now have motion on S^3, the unit hypersphere in four dimensions. This still holds true for Octonions, with unit norm still directly maintained when multiplying unit norm objects in general. Now the motion is that of a point on a seven dimensional hypersphere S^7. Sedenions are not normed division algebras, lacking linear alternativity and the moufang loop identities [17], thus multiplication of unit norm objects for sedenions (points on S^{15}) will not, generally, remain unit norm, i.e., will leave the S^{15} space.

The question then arises is there is a sub-algebra or projection in the sedenions, that is not just trivially the octonions, that can still allow unit norm propagation? If this works for Sedenions, what about Bi-sedenions (trigintaduonions) and higher dimensional Cayley algebras? In Ch. 2 (also [1]) it is shown that there are two Sedenion subspaces where the unit norm property is retained. This is found again at the level of the Bi-Sedenions by a similar construction. The results were initially explored computationally [1], then later established in theoretical proofs [1,6,15,16]. In those proofs a key step fails when attempting to go to higher orders beyond the bi-sedenions and its sub-algebra propagation.

In the RCHO(ST) hypothesis Physics unification was thought to directly entail propagation in terms of hypercomplex numbers [18] (from Reals thru Trigintaduonions in Cayley sequence). This hypothesis was motivated by Maxwell, Feynman and Cayley, in hopes of being able to directly encode the standard model and statistical mechanics. In the end, this idea was not ambitious enough, as will be clear in what follows. Part of the problem is that to get the 10D propagation formalism entails 'projections', not the more familiar mathematical objects directly giving rise to standard propagation (in a complex Hilbert space). Instead, the standard propagation is part of the emergent (with complex Hilbert space) description, as will described later.

The Feynman-Cayley Path Integral proposed in [1] involved use of chiral trigintaduonions in an effort to identify a mathematical framework within which to have a unified propagator theory (and maximal information propagation was sought for such a hypothesized propagator). At its root,

6

this is a hypothesis for an algebraic reality, with algebraic elements describing 'reality' and algebraic multiplicative processes underlying propagation. All of the different 'paths' of propagation are then brought together in a sum – where stationary phase is selected out and the variational calculus basis for much of physics then takes over to offer all of the familiar elegant solutions of classical physics. This is still thought to be the process, but two stages of emergence are indicated: (1) emergence of the emanation (projective) process followed by the (2) emergence of standard propagation in a complex Hilbert space. So, even though we start with RCHO(ST) with the emergence of *emanation*, we end with a framework for emergence of standard propagation with complex propagator in a complex Hilbert space.

With the Feynman-Cayley construction there is a sum on all algebras, with selection for the highest order unit norm propagating algebra. It is shown that the highest order propagating structure is the ten dimensional (10D) unit-norm trigintaduonions elements, that are used here, that are (chiral) extended sedenions that are themselves made from chiral extended octonions.

In quantum physics unitary propagation is a standard part of the description. Efforts to move to algebras to describe such propagation leads to formulations based on the normed division algebras (real, complex, quaternion, and octonion). In an effort to achieve maximal information propagation we relax the unitarity condition and show that multiplication (right) on a unit norm trigintaduonion base by a unit norm chiral trigintaduonions emanator results in a new unit norm product [1]. A path is comprised of repeated (right) multiplications. Each step of the 'emanation' arrived at is a multiplication by a 10D chiral trigintaduonion. Use of methods from noise budget analysis, a constructive perturbation analysis, as well as analysis relating to maximal perturbation according to the Kato Rellich theorem, show that the chiral trigintaduonion with maximal perturbation (outside the 10D into surrounding 32D) has magnitude α, precisely the fine structure constant. A relation between α and π results. Suppose repeated achiral emanation steps can be described as an iterative mapping, with unit-norm constraint resulting in a quadratic relation on components, we then expect the Feigenbaum universal bifurcation parameter, C_∞, to appear according to the number of independent dimensions in a chiral trigintaduonion emanation step and the precise form of the "emanator" construction. The number of effective dimensions is shown to be 29 plus a little more, and a relation between α,

π and C_∞ results that is in agreement with the choice of emanator examined in computational studies shown here. The computational studies with the emanator are also explored via "random walks" in the trigintaduonion space during emanation and to explore noise additivity effects. We will show component-level evolution that behaves like a random walk, with random walk asymptotics (established computationally). This will help to establish that the Emanation process is Martingale, since random walk processes are Martingale.

Just from the propagation structure on one path we already see core emergent structure that results in a universal emanation with structural parameters 10,22,78,137 and perturbation maximum α=~1/137. The central notion in the universal emanation hypothesis is that there should be *maximal information flow*, where this is accomplished by finding the highest theoretical dimensionality of unit-norm 'propagation', here called an emanation, which turns out to be 10, then add the maximal perturbation that still allows unit-norm propagation, where that perturbation is into the space the 10D motion is embedded in, here a 32 dimensional (trigintaduonion algebra) space.

1.4 Unit-norm propagation

For physical description a unit norm object can be used to represent a system, and by repeated transformation to other unit norm objects, it thereby evolves. Mathematical objects that can effect this 'transformation' simply by the rule of multiplication would be objects like division algebras, ideals, and what I'll simply call projections or emanations. In the universal propagator we have a unit norm trigintaduonion (32D) and perform a right multiplication with a chiral (10D) unit norm 'alpha-step' (defined by a max perturbation α into the 29 free dimensions, given by 32 minus one for each chiral choice, and one for the unit normalization overall). Consider multiplication of a given (starting) trigintaduonion from the right with a chiral bi-sedenion as a 'projection' through the (chiral) step indicated. The repeated application and repeated 'chiral steps' thereby arriving at a path describing a chiral propagation. The resulting universal propagation consists of a 32D unit norm trigintaduonion with propagation via right multiplication using a unit-norm, chiral bi-sedenion, with max-α perturbation.

We thereby arrive at a 'Universe Propagator' that takes on the physics parameters desired (notably the fine-structure constant) and imprints them onto the evolution as seen from the 'internal reference frame' where we reference an object in the 4D spacetime with Standard Model gauge field, and where the standard Lagrangian emerges as the necessary 'propagate-able' structure (where Hilbert space must be complex, not real, quaternionic or octonionic, etc. [19]). From maximum information flow with the constructs, and the required emergent complex Hilbert space (thus complex path integral, thus standard quantum operator formalism) we arrive back at the familiar results with justification of their core mathematical representations (e.g., complex Hilbert space), and now with justification of all parameters, all from the emanation hypothesis.

Unit-norm right product propagation is trivial for the division algebras since $norm(XY) = norm(X) \times norm(Y)$. From this it is apparent that we have an automorphism group given by the norm itself (since an automorphism if $A(XY)=A(X)A(Y)$), and in the case of the octonions this automorphism group is G2 [20]. It can be shown that SU(3) is in G2 [20]. Let's now consider the situation with a higher-order Cayley algebra, the Sedenions, 'S'. We obviously don't have $norm(S_1 S_2) = norm(S_1) \times norm(S_2)$ in general, as this would then allow S to join the ranks of the division algebras, and it is proven that such don't exist above the Octonions [21]. Can we still have a propagation structure? Is it possible to have a 'base' sedenion for which $norm(S_{base})=1$, and to have a right propagator (product) sedenion also $norm(S_{right})=1$, such that $norm(S_{base} \times S_{right}) =1$? The answer is yes (see appendix of [22] and [1]), when the sedenion has the (chiral) form of an octonion crossed with a real octonion: $S_{chiral} = (O,O_{real})$ or $S_{chiral} = (O_{real},O)$. Can we continue this to arrive at a propagation structure on the Trigintaduonions? Again the answer is yes, with the chiral form generalizing off the chiral Sedenion as might be expected: $T_{chiral} = (S_{chiral}, S_{real})$ or (S_{real}, S_{chiral}) [1]. It is proven that this extension process will go no further [1]. What happens is that due to the chiral form we are still able to re-express all T products (or S) as collections of terms involving tri-octonionic products (which have nice properties as described in [1]), and this can no longer occur above the (chiral) trigintaduonion level.

1.5 The Emanation Process
A 'deeper' phase of universal evolution is described by a theory of emanations, where mathematically invariant emergent structures appear:

emanation → propagation → trajectory

At the emanation to propagation emergence, one of the emergent constructs is the familiar path integral based on standard (unitary) propagators in a complex Hilbert space.

We have α, 10,22,78,137 as parameters resulting from analysis on a single path maximal information flow construct, where the number 22 corresponds to the number of emergent parameters in the description of the propagating construct (exact derivation of the 22-parameter to follow). In addition, the time choice is emergent via a multi-path construct, along with the propagator construct, and is coupled in both time step and imaginary time increment. The formulation is inherently embedded in a higher dimensional complex space, thus all of the QFT complex analysis analyticity methods are valid as the assumptions made are now part of the maximal information flow emergent construct.

1.6 Universal Thermality
There is a fundamental complex structure that still remains from the Emanator formalism upon projection from the higher-dimensional Cayley algebras into the maximum propagation described, for the alpha maximum-perturbation trigintaduonion. That complex structure is 'contact' analytic and is realized under conditions where limits involving that structure are taken to zero (or to some fixed value at component-level). Two main examples where this has been most significant are in (i) shifting a QFT to a thermal QFT by shifting to imaginary time related to the inverse temperature [66]; and (ii) in the use of dimensional regularization to renormalize QED and QCD [67]. Thus, the process of emanator theory settling on the maximum information propagation dimension is one where higher order complex structure (from non-propagate-able dimensions) is still accessible for regularization processes. A possible *origin* for the complex structures is described in a later section.

1.7 The maximum chiral emanation perturbation possible is alpha
The chiral trigintaduonion emanation described here gives a precise derivation for the mysterious physics constant α (the fine-structure constant) from the mathematical physics formalism providing maximal information propagation, with α being the maximal perturbation amount (a fractal limit), and π being the maximum amount of overall imaginary component contributing to that maximal perturbation. The maximal imaginary component is hypothesized to be at antiphase, thus 'π' phase angle. Component sums becoming angle sums is an aspect of the analyticity hypothesized for the Emanator theory (where exponential map

10

is used), and will be discussed in detail in what follows. Thus, α can be determined by a (fractal) limit process, and separately, by a maximal information propagation argument, where a relation can be shown to exist with the maximum antiphase amount 'π'. The ideal constructs of planar geometry, and related such via complex analysis, give methods for calculation of π to incredibly high precision (trillions of digits), thereby providing an indirect derivation of α to similar precision.

The trigintaduonion formulation provides the structure of the space for a deeper underlying information 'propagation' (with initial propagation being referred to as 'emanation'), with a precise derivation, and with a unit-norm perturbative limit that leads to an iterative-map-like computed α (a limit that is precisely related to the Feigenbaum bifurcation constant and thus fractal). The familiar Mandelbrot set: $f(z) = z^2 + c$ (complex) has C_∞ as a limit value on c in the iterative map for stability. Similarly, and directly relevant here, for the real one-parameter map $f(x) = a - x^2$, C is the (universal) limit value on a [2]. Since repeated chiral emanation steps can be described as an iterative mapping, with unit-norm constraint resulting in a quadratic relation on components, we expect the Feigenbaum universal bifurcation parameter, C_∞, to appear according to the number of independent dimension in a chiral trigintaduonion emanation step and the precise form of the "emanator" construction. The number of effective dimensions is shown to be 29 plus a little more, where a relation between α, π and C_∞ results that is dependent on the choice of (achiral) emanator. The emanator identified by the $\{\alpha, \pi\ C_\infty\}$ relation is used to explore random walks in the Trigintaduonion space during emanation and explore noise additivity effects.

It is hypothesized that standard model physics with path integral propagation, choice of time, and its assortment of fundamental constants, is emergent from maximal information emanation via trigintaduonions and this will be explored in what follows. Thus, in Emanator Theory, the form of quantum propagation is itself emergent, and within that construct, there is then emergent the functional optimization that describes how the system behaves, e.g., the Lagrangian and choice of time is part of that latter emergent step. Thus, Lagrangians originally introduced as a convenient mathematical constructs, and in later physics endowed with their own physicality, especially in conjunction with the path-integral description to properly capture topological features (the Aharanov-Bohm experiments), are here seen as direct mathematical encapsulations of the fundamental emergent nature of the physical system.

11

1.8 Standard Model

The chiral trigintaduonions T, with right product operation $((T \times T) \times T)\ldots$, used for maximum information transmission, is shown to be H×O when arranged for achiral emanation. When considering a sum over chiral emanations to obtain an achiral emanator, with T as phase factor, we have the exponentiation operation $\exp(iT)$, which leads to a theory that is $C \times H \times O$. As such, we have the foundation for the associative operator algebra of the Standard Model: $U(1) \times SU(2)_L \times SU(3)$ [23]. A complication with T products is you can have zero divisors. A framework is adopted to remove the zero divisors by requirement of maximum domain of analyticity on the log trigintaduonion multiplication, resulting in a description for the meromorphic precipitation of matter. In this process a fundamental quantum is indicated from the zero-divisor residue terms. Analyticity in the form of a Wick rotation also provides a mechanism whereby we can transition to a dimensionful action and quantum and arrive at an explanation for the critical 'smallness' of Planck's constant. A review of emanation theory will be given first, including the origin of α, the fine-structure constant, followed by showing that the form of the emanator is $T_{em} \cong H \times O$, followed by a description of the possible meromorphic origin of point-like matter.

Getting an associative algebra from the repeated operation of a non-associative algebra is first described in [20] in the context of repeated octonion products: $((O \times O) \times O)\ldots$, where the algebra SU(3) can result. This is found to be equivalent to fixing one of the octonion imaginary components in such a repeated-product operation [20,23]. Dixon shows in [23] that the $C \times H \times O$ product algebra lays the foundation for the associative operator algebra of the Standard Model: $U(1) \times SU(2)_L \times SU(3)$. In later work this is explored in the form of ideals [24].

Chapter 2. The Maximum Information Emanation Hypothesis and the 10D unit-norm emanation structure

2.1 Hypercomplex Numbers

Physics has a lengthy 'love-hate' relationship with hypercomplex numbers. One of the earliest formulations of electromagnetism by Maxwell involved quaternionic mathematics, and even at that time this relationship was off to a difficult start. As stated by Maxwell in a manuscript on the application to electromagnetism in November of 1870 [25]: "... The invention of the Calculus of Quaternions by Hamilton is a step towards the knowledge of quantities related to space which can only be compared for its importance with the invention of triple coordinates by Descartes. The limited use which has up to the present time been made of Quaternions *must be attributed partly to the repugnance of most mature minds to new methods involving the expenditure of thought ...*" (with emphasis mine). The enthusiasm of Maxwell for use of Quaternionic mathematics did not win over other great physicists of his day, such as Josiah Willard Gibbs and Oliver Heaviside, who discarded the quaternionic mathematics in favor of a new mathematics (vector calculus) that they invented so as to avoid the 'foreign' hypercomplex mathematics. In a biography of Hamilton [26], in a quotation attributed to Gibbs: "My first acquaintance with quaternions was in reading Maxwell's E.&M. where Quaternion notations are considerably used. ... I saw, that although the methods were called quaternionic the idea of the quaternion was quite foreign to the subject."

The stigma associated with hypercomplex mathematics, and the higher-dimensional physics unification attempts of Maxwell and later Einstein, was still significant decades later when Feynman obtained an unusual proof of the homogeneous Maxwell equations [27-30] in a higher (than 3) dimensional space. Feynman was trying to see if any new theoretical theory would be indicated and the fact that he had obtained a novel new way to explain the existing Maxwell's equations in higher dimensions was not interesting at the time. The inextricable problems of quantum gravity and the discovery of higher-dimensional string theory, among other things, have changed the focus since that time.

13

It has been shown in numerous papers that the (1,9) dimensional superstring has a natural parameterization in terms of octonions [31-33]. In [8,9] the Dirac and Maxwell equations (in vacuum) are derived using octonionic algebras. In [10] a quaternionic equation is described for electromagnetic fields in inhomogeneous media. In [11], the D4-D5-E6 model that includes the Standard Model plus Gravity is constructed using octonionic fermion creators and annihilators. In [12] octonionic constructions are shown to be consistent with the SU(3)$_C$ gauge symmetry of QCD. It would appear that there are a number of implementations involving hypercomplex numbers that are consistent with the Standard Model. But there is still the question of why bother? What is shown here is why the bother might be worth it as the new mathematical formulations recovers simultaneously the quantum theory; the critical constant alpha; and many key relations, such as manifest Lorentz Invariance.

2.2 The Cayley Algebras

The list representation for hypercomplex numbers will make things clearer in what follows so will be introduced here for the first seven Cayley algebras:

Reals: X_0 → (X_0) .
Complex: $(X_0 + X_1 i)$ → (X_0, X_1) .
Quaternions: $(X_0 + X_1 i + X_2 j + X_3 k)$ → (X_0, X_1, X_2, X_3) → $(X_0, \ldots, X_3)$.
Octonions: $(X_0, \ldots, X_7)$ with seven imaginary numbers.
Sedenions: $(X_0, \ldots, X_{15})$ with fifteen imaginary numbers.
Trigintaduonions (a.k.a Bi-Sedenions): $(X_0, \ldots, X_{31})$ with 31 imaginary numbers.
Bi-Trigintaduonions: $(X_0, \ldots, X_{63})$ with 63 types of imaginary number.

Consider how the familiar complex numbers can be generated from two real numbers with the introduction of a single imaginary number 'i', $\{X_0, X_1\}$ → $(X_0 + X_1 i)$. This construction process can be iterated, using two complex numbers, $\{Z_0, Z_1\}$, and a new imaginary number 'j':

$$(Z_0 + Z_1 j) = (A+Bi) + (C+Di)j = A+Bi + Cj +Dij = A+Bi + Cj +Dk,$$

where we have introduced a third imaginary number 'k' where '$ij=k$'. In list notation this appears as the simple rule ((A,B),(C,D)) = (A,B,C,D). This iterative construction process can be repeated, generating algebras

14

doubling in dimensionality at each iteration, to generate the 1,2,4,8,16, 32, and 64 dimensional algebras listed above. The process continues indefinitely to higher orders beyond that, doubling in dimension at each iteration, but we will see that the main algebras of interest for physics are those with dimension 1,2,4,and 8, and sub-spaces of those with dimension 16 and 32 dimensional algebras.

Addition of hypercomplex numbers is done component-wise, so is straightforward. For hypercomplex multiplication, list notation makes the freedom for group splittings more apparent, where any hypercomplex product ZxQ to be expressed as (U,V)x(R,S) by splitting Z=(U,V) and Q=(R,S). This is important because the product rule, generalized by Cayley, uses the splitting capability. The Cayley algebra multiplication rule is:

$$(A,B)(C,D) = ([AC-D*B],[BC*+DA]),$$

where conjugation of a hypercomplex number flips the signs of all of its imaginary components:

$$(A,B)* = Conj(A,B) = (A*,-B)$$

The specification of new algebras, with addition and multiplication rules as indicated by the constructive process above, is known as the Cayley-Dickson construction, and this gives rise to what is referred to as the Cayley algebras in what follows.

If a Split Cayley algebra is used, then the multiplication rule has a single sign difference:

$$(A,B)(C,D) = ([AC+D*B],[BC*+DA]).$$

If you use the Cayley-Dickson procedure to double the octonions to get the sedenions, you retain the properties common to all Cayley-Dickson algebras [34]:

- centrality: if $xy = yx$ for all y in the algebra A, then x is in the base field of A, which is the real numbers R;
- simplicity: no ideal K other than {0} and the algebra A, or, equivalently, if for all x in K and for all y in A xy and yx are in K, then K = {0} or A;
- flexibility: $(x,y,z) = (xy)z - x(yz) = -(z,y,x)$, or, equivalently, $(xy)x = x(yx) = xyx$;

15

- power-associativity: (xx)x = x(xx) and ((xx)x)x = (xx)(xx), or, equivalently, $x^{\wedge}m\ x^{\wedge}n = x^{\wedge}(m+n)$;
- Jordan-admissibility: xoy = (1/2)(xy + yx) makes a Jordan algebra;
- degree two: xx - t(x)x + n(x) = 0, for some real numbers t(x) and n(x);
- derivation algebra G2 for octonions and beyond; and
- squares of basic units = -1.

For sedenions, you lose the following properties:
(1) the division algebra (over R) property xy = 0 only if x ≠ 0 and y ≠ 0. A concrete example of zero divisors in terms of that basis is given by [35]: (e1 + e10)(e15 - e4) = -e14 - e5 + e5 + e14 = 0.).
(2) linear alternativity: (x,y,z) = (xy)z - x(yz) = (-1)P(Px,Py,Pz), where P is a permutation of sign (-1)P; and
(3) the Moufang identities: (xy)(zx) = x(yz)x; (xyx)z = x(y(xz)); z(xyx) = ((zx)y)x.

For sedenions, you retain the following properties:
(1) anticommutativity of basic units: xy = -yx; and
(2) nonlinear alternativity of basic units: (xx)y = x(xy) and (xy)y = x(yy).

2.3 Chiral Extension from the Octonions to the Chiral Sedenions

Further theoretical details on hypercomplex numbers can be found at [17,36]. In what follows multiplications involving unit norm Cayley numbers will be done at the various orders using the Cayley algebra multiplication rule described above, that reduces the order of hypercomplex complex multiplication, which when iterated allows all hypercomplex products to reduce to a collection of Real multiplications. Millions of repeated hypercomplex multiplications are done computationally to demonstrate unit norm propagation in the situations that follow, where B denotes a bisedenion, S denotes a sedenion, O a octonion, Q a quaternion, C for complex, and R for a real:

Sedenions have two unit norm propagators of the form:

S(unit norm) × S(unit norm propagator) = S(unit norm)
S(unit norm) = $S_1(O_{Left}, O_{Right}) = S_1(O_L, O_R) = (O_{1L}, O_{1R})$

16

If S_1 is unit norm, then $norm(S_1) = S_1 \times S_1{}^* = 1$, which for our notation means:

$1 = (O_{1L}, O_{1R}) \times (O_{1L}{}^*, -O_{1R}) = ([O_{1L} \times O_{1L}{}^* + O_{1R}{}^* \times O_{1R}],$
$[-O_{1R} \times O_{1L} + O_{1R} \times O_{1L}])$
$1 = ([norm(O_{1L}) + norm(O_{1R})], 0)$
$1 = norm(O_{1L}) + norm(O_{1R})$

S(unit norm propagator) $= S_2(O_{Left}, O_{Real}) = (O_{2L}, \alpha)$ for the right octonion real, e.g., in list notation have $O_{Real} = (\alpha, 0, 0, 0, 0, 0, 0, 0)$, so have $(O_{2L}, (\alpha, 0, 0, 0, 0, 0, 0, 0))$ which is abbreviated as (O_{2L}, α) where it is understood that α is real and is the real part of the purely real right octonion. There is another type of unit norm propagator where we have (O_{Real}, O_{Right}) where the same results hold, but the example that follows will use the (O_{2L}, α) form.

If S_2 is unit norm, then $norm(S_2) = S_2 \times S_2{}^* = 1$, which for our notation means: $1 = norm(O_{2L}) + \alpha^2$.

So we can now ask the question, Does S(unit norm) × S(unit norm propagator), return a unit norm Sedenion when using the special class of unit norm propagators indicated?

Proof that Norm(S₁×S₂)=1

$(S_1 \times S_2) = (O_{1L}, O_{1R}) \times (O_{2L}, \alpha) = ([O_{1L} \times O_{2L} - \alpha O_{1R}], [\alpha O_{1L} + O_{1R} \times O_{2L}{}^*])$
$(S_1 \times S_2)^* = ([O_{1L} \times O_{2L} - \alpha O_{1R}]^*, -[\alpha O_{1L} + O_{1R} \times O_{2L}{}^*])$

$norm(S_1 \times S_2) = (S_1 \times S_2) \times (S_1 \times S_2)^*$
$= ([O_{1L} \times O_{2L} - \alpha O_{1R}] \times [O_{1L} \times O_{2L} - \alpha O_{1R}]^* + [\alpha O_{1L} + O_{1R} \times O_{2L}{}^*]^* \times [\alpha O_{1L} + O_{1R} \times O_{2L}{}^*],$
$-[\alpha O_{1L} + O_{1R} \times O_{2L}{}^*] \times [O_{1L} \times O_{2L} - \alpha O_{1R}] + [\alpha O_{1L} + O_{1R} \times O_{2L}{}^*] \times [O_{1L} \times O_{2L} - \alpha O_{1R}])$

$= (norm(O_{1L} \times O_{2L}) + norm(O_{1R} \times O_{2L}{}^*) + \alpha^2 norm(O_{1R}) + \alpha^2 norm(O_{1L})$
$-\alpha(O_{1L} \times O_{2L}) \times O_{1R}{}^* - \alpha O_{1R} \times (O_{1L} \times O_{2L})^* + \alpha O_{1L}{}^* \times (O_{1R} \times O_{2L}{}^*) + \alpha(O_{1R} \times O_{2L}{}^*)^* \times O_{1L}, \mathbf{0})$

Multiplying the expressions previously obtained, $1 = norm(O_{1L}) + norm(O_{1R})$ with $1 = norm(O_{2L}) + \alpha^2$, and making use of the norm property $norm(xy) = norm(x) norm(y)$, we have:

$$\text{norm}(S_1 \times S_2) = (1-\alpha Z, 0), \text{ where,}$$
$$Z = +(O_{1L} \times O_{2L}) \times O_{1R}{}^* + O_{1R} \times (O_{1L} \times O_{2L})^*$$
$$-O_{1L}{}^* \times (O_{1R} \times O_{2L}{}^*) - (O_{1R} \times O_{2L}{}^*)^* \times O_{1L}.$$

Since we are computing the norm, which returns only the real component, we know Z must be real. To work with this expression with a little more clarity, switch to the notation:

$A=O_{1L}$; $B=O_{2L}$; $C=O_{1R}{}^*$, then have
$$Z = (A \times B) \times C + C^* \times (A \times B)^* - A^* \times (C^* \times B^*) - (C^* \times B^*)^* \times A$$
$$Z = (A \times B) \times C + C^* \times (A \times B)^* - A^* \times (B \times C)^* - (B \times C) \times A$$

The Cayley algebras up to octionic are also known as the composition algebras for which a number of properties exist. We need the braid laws to proceed, so let's briefly detour to address that. The fundamental composition rule is simply that of the norm of a product being the product of the norms: norm(XY) = norm(X) x normY) Consider the norm of two things added:

$$\text{Norm}(X+Y) = (X+Y)(X+Y)^* = XX^* + XY^* + YX^* + YY^*$$
$$= \text{norm}(X) + \text{norm}(Y) + 2\ \text{real}(XY^*)$$

Define $[X,Y] = \text{real}(XY^*) = [\text{norm}(X+Y) - \text{norm}(X) - \text{norm}(Y)]/2$, then have another way to express conjugation using norms and real parts:

$X^* = 2[X,1] - X = 2\text{real}(X) - X = (\text{real}(X) \text{ unchanged, imag}(X) \text{ negated}))$

The composition algebras (up to octionic) build from the core norm(XY) = norm(X) x normY) relation to arrive at a number of interesting properties, including the 'braid' laws: $[XY,Z] = [Y,X^*Z]$ and $[XY,Z]=[X,ZY^*]$. To arrive at the Braid law (following [34]) you start with the composition law norm(XY)=norm(X)norm(Y), you then prove the scaling law, $[XY,XZ]=\text{norm}(x)[Y,Z]$, by substituting Y with Y+Z in the composition law. Then establish the exchange law $[XY,UZ] = 2[X,U][Y,Z]-[XZ,UY]$ by substituting X with X+U in the scaling law. If you put U=1 in the exchange law, it reduces to forms allowing the braid law to be shown.

Let's apply the braid law for the form [XY,Z] to the (B×C)×A term, so let's look at the braid law for $[BC,A^*] = [C,B^*A^*]$, which can be rewritten as:

$\text{norm}(BC+A^*)-\text{norm}(BC)-\text{norm}(A^*) =$
$\text{norm}(C+B^*A^*)-\text{norm}(C)-\text{norm}(B^*A^*)$
$\text{norm}(BC+A^*)=\text{norm}(BC)+\text{norm}(A^*)+(BC)A+A^*(BC)^*$
$\text{norm}(C+B^*A^*)=\text{norm}(C)+\text{norm}(B^*A^*)+C(AB)+(AB)^*C^*$

putting this together: $(BC)A+A^*(BC)^*=C(AB)+(AB)^*C^*$. So we can now rewrite the $(B{\times}C){\times}A$ term as: $(B{\times}C){\times}A = C{\times}(A{\times}B)+(A{\times}B)^*{\times}C^*-A^*{\times}(B{\times}C)^*$. Substituting this back into Z:

$$Z = (A{\times}B){\times}C + C^*{\times}(A{\times}B)^*-C{\times}(A{\times}B)-(A{\times}B)^*{\times}C^*$$
$$= [(A{\times}B){\times}C-C{\times}(A{\times}B)] + [C^*{\times}(A{\times}B)^*-(A{\times}B)^*{\times}C^*]$$

What is a commutator on the Cayley numbers, is it necessarily non-real?

$$XY=(A,B)(C,D) = ([AC-D^*B],[BC^*+DA])$$
$$YX=(C,D)(A,B)= ([CA-B^*D],[DA^*+BC])$$
$$\{X,Y\}=XY-YX= ([AC-CA+B^*D-D^*B],[BC^*-BC+DA-DA^*])$$
$$\{X,Y\}= (\ [\{A,C\}+2\text{Im}(B^*D)],\ [B\ 2\text{Im}(C) + D\ 2\text{Im}(A)]\ )$$

So the commutator at one order of Cayley number is reduced to an expression involving the commutator at the next lower order Cayley number, plus a bunch of other terms that don't contribute to the real component. This can be iterated to arrive at the real algebra in the commutator, where the commutator is zero, thereby establishing that the commutator on the Cayley numbers must result in a pure imaginary Cayley number. This being the case, we see that since Z consists of two commutator terms, neither of which has a real contribution, and since Z must be real, this proves that Z=0.

This proves the first extension, for unit-norm propagators that are Sedenions of the form $S_{\text{Left}}=(O_{\text{Left}},\alpha)$ or $S_{\text{Right}}=(\alpha,O_{\text{Right}})$, where O_{Left} and O_{Right} are any octonion. The next extension is to unit-norm propagators that are Bisedenion by using similar constructions, e.g., Bisedenions, of the form $B=(S_{\text{Left}},S_{\text{Real}}) = (\ (O_{\text{Left}},\alpha)\ ,\ \beta\ )$. (Note that α is a real octonion, while β is a purely real sedenion.)

2.4 Chiral Extension from the Chiral Sedenions to the Chiral Trigintaduonions

Bisedenions have two unit norm propagators of the form:
$B(\text{unit norm}) \times B(\text{unit norm propagator}) = B(\text{unit norm})$
$B(\text{unit norm}) = B_1(S_{\text{Left}},S_{\text{Right}}) = B_1(S_L,S_R) = (S_{1L},S_{1R})$

If B_1 is unit norm, then norm$(B_1) = B_1 \times B_1^* = 1$, which for our notation means:

$1 = (S_{1L}, S_{1R}) \times (S_{1L}^*, -S_{1R}) =$

$([S_{1L} \times S_{1L}^* + S_{1R}^* \times S_{1R}], [-S_{1R} \times S_{1L} + S_{1R} \times S_{1L}])$

$1 = ([\text{norm}(S_{1L}) + \text{norm}(S_{1R})], 0)$

$1 = \text{norm}(S_{1L}) + \text{norm}(S_{1R})$

B(unit norm propagator) = $B_2(S_{\text{Left}}, S_{\text{Real}})$ = (S_{2L}, β) for the right sedenion real, e.g., in list notation have S_{Real} = $(\beta,0,0,0,0,0,0,0,0,0,0,0,0,0,0,0)$, so have $(O_{2L}, (\beta,0,0,0,0,0,0,0,0,0,0,0,0,0,0,0))$ which is abbreviated as (O_{2L}, β) where it is understood that β is real and is the real part of the purely real right sedenion. There is another type of unit norm propagator where we have $(S_{\text{Real}}, S_{\text{Right}})$ where the same results hold, but the example that follows will use the (S_{2L}, β) form.

If B_2 is unit norm, then norm$(B_2) = B_2 \times B_2^* = 1$, which for our notation means: $1 = \text{norm}(S_{2L}) + \beta^2$.

So we can now ask the question:
Does B(unit norm) $\times$ B(unit norm propagator), return a unit norm Bisedenion when using the special class of unit norm propagators indicated?

Proof that Norm$(B_1 \times B_2)=1$

$(B_1 \times B_2) = (S_{1L}, S_{1R}) \times (S_{2L}, \beta) = ([S_{1L} \times S_{2L} - \beta S_{1R}], [\beta S_{1L} + S_{1R} \times S_{2L}^*])$

$(B_1 \times B_2)^* = ([S_{1L} \times S_{2L} - \beta S_{1R}]^*, -[\beta S_{1L} + S_{1R} \times S_{2L}^*])$

norm$(B_1 \times B_2) = (B_1 \times B_2) \times (B_1 \times B_2)^*$

$= ([S_{1L} \times S_{2L} - \beta S_{1R}] \times [S_{1L} \times S_{2L} - \beta S_{1R}]^* + [\beta S_{1L} + S_{1R} \times S_{2L}^*]^* \times [\beta S_{1L} + S_{1R} \times S_{2L}^*],$

$-[\beta S_{1L} + S_{1R} \times S_{2L}^*] \times [S_{1L} \times S_{2L} - \beta S_{1R}] + [\beta S_{1L} + S_{1R} \times S_{2L}^*] \times [S_{1L} \times S_{2L} - \beta S_{1R}])$

$= (\text{norm}(S_{1L} \times S_{2L}) + \text{norm}(S_{1R} \times S_{2L}^*) + \beta^2 \text{norm}(S_{1R}) + \beta^2 \text{norm}(S_{1L})$

$-\beta(S_{1L} \times S_{2L}) \times S_{1R}^* - \beta S_{1R} \times (S_{1L} \times S_{2L})^* + \beta S_{1L}^* \times (S_{1R} \times S_{2L}^*)$

$+\beta(S_{1R} \times S_{2L}^*)^* \times S_{1L}, 0)$

20

To proceed as before we need to show that the norm property norm(xy) = norm(x)norm(y) holds for the sedenions when one of them is constrained to be in the form of the sedenion propagator, e.g., does norm($S_{1L} \times S_{2L}$)=norm(S_{1L})×norm(S_{2L}) where S_{2L} is in the form of the sedenion propagator?

norm($S_{1L} \times S_{2L}$) = ($S_{1L} \times S_{2L}$) × ($S_{1L} \times S_{2L}$)*
= ([$O_{1LL} \times O_{2LL} - \alpha O_{1LR}$] × [$O_{1LL} \times O_{2LL} - \alpha O_{1LR}$]* +
  [$\alpha O_{1LL} + O_{1LR} \times O_{2LL}$*]* × [$\alpha O_{1LL} + O_{1LR} \times O_{2LL}$*],
  −[$\alpha O_{1LL} + O_{1LR} \times O_{2LL}$*] × [$O_{1LL} \times O_{2LL} - \alpha O_{1LR}$] +
  [$\alpha O_{1LL} + O_{1LR} \times O_{2LL}$*] × [$O_{1LL} \times O_{2LL} - \alpha O_{1LR}$])

= (norm($O_{1LL} \times O_{2LL}$)+norm($O_{1LR} \times O_{2LL}$*)+α^2 norm(O_{1LR})+α^2 norm(O_{1LL})−α($O_{1LL} \times O_{2LL}$)×O_{1LR}*−αO_{1LR}×($O_{1LL} \times O_{2LL}$)*+ αO_{1LL}*×($O_{1LR} \times O_{2LL}$*)+α($O_{1LR} \times O_{2LL}$*)*×O_{1LL}, **0**)

Now that we've reduced to this level, we know that the octonions will offer the standard norm property whereby norm($O_{1LL} \times O_{2LL}$)=norm(O_{1LL})norm(O_{2LL}) and we show the other terms are zero since real yet consisting of commutators, the latter arrangements made possible by manipulations according to the braid laws that hold for the composition algebras (including the octonions) without restriction.

So as before, by multiplying the expressions previously obtained, 1 = norm(S_{1L}) + norm(S_{1R}) with 1 = norm(S_{2L})+β^2, and making use of the norm property norm(xy)=norm(x)norm(y) applicable for the terms of interest, we have:

$$norm(B_1 \times B_2) = (1-\beta Z, 0), \text{ where,}$$
$$Z = +(S_{1L} \times S_{2L}) \times S_{1R}* + S_{1R} \times (S_{1L} \times S_{2L})*$$
$$-S_{1L}* \times (S_{1R} \times S_{2L}*) - (S_{1R} \times S_{2L}*)* \times S_{1L}.$$

Since we are computing the norm, which returns only the real component, we know Z must be real. As with the lower order Cayley extension, we need the braid laws to proceed at this juncture.

What is ($S_{1L} \times S_{2L}$)×S_{1R}* when accounting for the special form of S_{2L}=(O_{2LL}, α)? First calculate ($S_{1L} \times S_{2L}$):

($S_{1L} \times S_{2L}$)=(O_{1LL}, O_{1LR})(O_{2LL}, α) = ([$O_{1LL} \times O_{2LL} - \alpha O_{1LR}$], [$\alpha O_{1LL} + O_{1LR} \times O_{2LL}$*])

21

Then

$(S_{1L} \times S_{2L}) \times S_{1R}^* = ([O_{1LL} \times O_{2LL} - \alpha O_{1LR}], [\alpha O_{1LL} + O_{1LR} \times O_{2LL}^*]) (O_{1RL}^*, -O_{1RR})$

$= ([O_{1LL} \times O_{2LL} - \alpha O_{1LR}] O_{1RL}^* + O_{1RR}^* [\alpha O_{1LL} + O_{1LR} \times O_{2LL}^*],$

$\quad -O_{1RR} [O_{1LL} \times O_{2LL} - \alpha O_{1LR}] + [\alpha O_{1LL} + O_{1LR} \times O_{2LL}^*] O_{1RL})$

$= ((O_{1LL} \times O_{2LL}) \times O_{1RL}^* - \alpha O_{1LR} \times O_{1RL}^* + \alpha O_{1RR}^* \times O_{1LL} + O_{1RR}^* \times (O_{1LR} \times O_{2LL}^*),$

$\quad -O_{1RR} \times (O_{1LL} \times O_{2LL}) + \alpha O_{1RR} \times O_{1LR} + \alpha O_{1LL} \times O_{1RL} + (O_{1LR} \times O_{2LL}^*) \times O_{1RL})$

$S_{1R} \times (S_{1L} \times S_{2L})^* = (O_{1RL}, O_{1RR}) ([O_{1LL} \times O_{2LL} - \alpha O_{1LR}]^*,$

$\quad -[\alpha O_{1LL} + O_{1LR} \times O_{2LL}^*])$

$= (O_{1RL} \times [(O_{1LL} \times O_{2LL})^* - \alpha O_{1LR}^*] + [\alpha O_{1LL}^* + (O_{1LR} \times O_{2LL}^*)^*] \times O_{1RR},$

$\quad -[\alpha O_{1LL} + O_{1LR} \times O_{2LL}^*] \times O_{1RL} + O_{1RR} \times [O_{1LL} \times O_{2LL} - \alpha O_{1LR}])$

$= (O_{1RL} \times (O_{1LL} \times O_{2LL})^* - \alpha O_{1RL} \times O_{1LR}^*$

$+ \alpha O_{1LL}^* \times O_{1RR} + (O_{1LR} \times O_{2LL}^*)^* \times O_{1RR},$

$\quad -\alpha O_{1LL} \times O_{1RL} - (O_{1LR} \times O_{2LL}^*) \times O_{1RL} + O_{1RR} \times (O_{1LL} \times O_{2LL}) - \alpha O_{1RR} \times O_{1LR})$

Putting these first two terms together:

$+(S_{1L} \times S_{2L}) \times S_{1R}^* + S_{1R} \times (S_{1L} \times S_{2L})^* =$

$((O_{1LL} \times O_{2LL}) \times O_{1RL}^* + O_{1RL} \times (O_{1LL} \times O_{2LL})^*$

$- \alpha O_{1LR} \times O_{1RL}^* + \alpha O_{1RR}^* \times O_{1LL} - \alpha O_{1RL} \times O_{1LR}^* + \alpha O_{1LL}^* \times O_{1RR}$

$+ O_{1RR}^* \times (O_{1LR} \times O_{2LL}^*) + (O_{1LR} \times O_{2LL}^*)^* \times O_{1RR}, \qquad 0)$

For $S_{1L}^* \times (S_{1R} \times S_{2L}^*)$ we have:

$(S_{1R} \times S_{2L}^*) = (O_{1RL}, O_{1RR})(O_{2LL}^*, -\alpha) = ([O_{1RL} \times O_{2LL}^* + \alpha O_{1RR}],$

$[-\alpha O_{1RL} + O_{1RR} \times O_{2LL}])$

So, $S_{1L}^* \times (S_{1R} \times S_{2L}^*) = (O_{1LL}^*, -O_{1LR}) \times ([O_{1RL} \times O_{2LL}^* + \alpha O_{1RR}],$

$[-\alpha O_{1RL} + O_{1RR} \times O_{2LL}])$

$= (O_{1LL}^* \times (O_{1RL} \times O_{2LL}^*) + \alpha O_{1LL}^* \times O_{1RR} -$

$\alpha O_{1RL}^* \times O_{1LR} + (O_{1RR} \times O_{2LL})^* \times O_{1LR} , \text{ term})$

While for $(S_{1R} \times S_{2L}^*)^* \times S_{1L}$ have

$(S_{1R} \times S_{2L}^*)^* \times S_{1L} =$

$([O_{1RL} \times O_{2LL}^* + \alpha O_{1RR}]^*, [\alpha O_{1RL} - O_{1RR} \times O_{2LL}]) \times (O_{1LL}, O_{1LR})$

$= ([O_{1RL} \times O_{2LL}^* + \alpha O_{1RR}]^* \times O_{1LL} - O_{1LR}^* \times [\alpha O_{1RL} - O_{1RR} \times O_{2LL}], \text{ - term})$

$S_{1L}^* \times (S_{1R} \times S_{2L}^*) + (S_{1R} \times S_{2L}^*)^* \times S_{1L} =$

$(O_{1LL}^* \times (O_{1RL} \times O_{2LL}^*) + (O_{1RL} \times O_{2LL}^*)^* \times O_{1LL}$

$+ \alpha O_{1LL}^* \times O_{1RR} - \alpha O_{1RL}^* \times O_{1LR} + \alpha O_{1RR}^* \times O_{1LL} - \alpha O_{1LR}^* \times O_{1RL}$

$+(O_{1RR}\times O_{2LL})^*\times O_{1LR}+O_{1LR}^*\times(O_{1RR}\times O_{2LL}), \quad 0)$

So have,

$$Z = (\quad \{(O_{1LL}\times O_{2LL})\times O_{1RL}^*+O_{1RL}\times(O_{1LL}\times O_{2LL})^*$$
$$-O_{1LL}^*\times(O_{1RL}\times O_{2LL}^*)-(O_{1RL}\times O_{2LL}^*)^*\times O_{1LL}\} +$$
$$\{O_{1RR}^*\times(O_{1LR}\times O_{2LL}^*)+(O_{1LR}\times O_{2LL}^*)^*\times O_{1RR}$$
$$-(O_{1RR}\times O_{2LL})^*\times O_{1LR}-O_{1LR}^*\times(O_{1RR}\times O_{2LL})\} +$$
$$-\alpha O_{1LR}\times O_{1RL}^*+\alpha\, O_{1RR}^*\times O_{1LL}-\alpha O_{1RL}\times O_{1LR}^* +\alpha O_{1LL}^*\times O_{1RR}$$
$$-\alpha O_{1LL}^*\times O_{1RR}+\alpha O_{1RL}^*\times O_{1LR}-\alpha O_{1RR}^*\times O_{1LL}+\alpha O_{1LR}^*\times O_{1RL} \ , 0)$$

$$Z = (\{im\}+\alpha\{O_{1RL}^*\times O_{1LR}+\alpha O_{1LR}^*\times O_{1RL}$$
$$-\alpha O_{1LR}\times O_{1RL}^*-\alpha O_{1RL}\times O_{1LR}^*\} \ , 0)$$
$$Z = (\{im\}+\alpha\{2Im\{O_{1RL}^*\times O_{1LR}\} +2Im\{O_{1LR}^*\times O_{1RL}\} \ , 0)$$

So again, have that Z = pure imaginary, and since it must be real, it is thus zero. Thus, we have norm($B_1\times B_2$) = 1. This proves the second extension, for unit-norm propagators that are Bisedenions of the form $B_{Left}=(S_{Left},\beta)$ or $B_{Right}=(\beta,S_{Right})$, where S_{Left} and S_{Right} are sedenion propagators shown in the first extension, e.g., $S_{Left}=(O_{Left},\alpha)$. (Note that α is a purely real octonion, while β is a purely real sedenion.)

2.5 Chiral extension of Chiral Trigintaduonion Fails
After the bisedenions (also known as trigintaduonions) come the bitrigintaduonions, the 64-component Cayley algebra (denoted by 'T' in following but later when I reference the RCHO(ST) hypothesis, the 'T' refers to trigintaduonions). Let's try extending further to see if we can have norm($T_1\times T_2$) = 1, when we build with a similar extension method to define our unit-norm propagator: $T_{Left}=(B_{Left},\gamma)$, $B_{Left}=(S_{Left},\beta)$, and $S_{Left}=(O_{Left},\alpha)$, where, as before, once we get to the octionic Cayley level we are unrestricted (e.g., O_{Left} can be any octonion). Let's see if we can construct, as before, a T unit norm propagators of the form:

T(unit norm) $\times$ T(unit norm propagator) = T(unit norm)
T(unit norm) = $T_1(B_{Left},B_{Right}) = T_1(B_L,B_R) = (B_{1L},B_{1R})$

If T_1 is unit norm, then norm(T_1) = $T_1 \times T_1^*$ = 1, which for our notation means:
$1 = (B_{1L},B_{1R})\times(B_{1L}^*,-B_{1R}) = ([B_{1L}\times B_{1L}^*+B_{1R}^*\times B_{1R}],$
$[-B_{1R}\times B_{1L}+B_{1R}\times B_{1L}])$
$1 = ([norm(B_{1L})+norm(B_{1R})], 0)$

$1 = \text{norm}(B_{1L}) + \text{norm}(B_{1R})$

T(unit norm propagator) $= T_2(B_{Left}, B_{Real}) = (B_{2L}, \gamma)$ for the right bisedenion real γ is real and is the real part of the purely real right bisedenion.

If T_2 is unit norm, then $\text{norm}(T_2) = T_2 \times T_2^* = 1$, which for our notation means: $1 = \text{norm}(B_{2L}) + \gamma^2$.

So we can now ask the question, Does T(unit norm) $\times$ T(unit norm propagator), return a unit norm bitrigintaduonion when using the special class of unit norm propagators indicated?

Failure of Proof construction for Norm($T_1 \times T_2$)=1 , and computational proof of failure of Norm($T_1 \times T_2$)=1

$(T_1 \times T_2) = (B_{1L}, B_{1R}) \times (B_{2L}, \gamma) = ([B_{1L} \times B_{2L} - \gamma B_{1R}], [\gamma B_{1L} + B_{1R} \times B_{2L}^*])$
$(T_1 \times T_2)^* = ([B_{1L} \times B_{2L} - \gamma B_{1R}]^*, -[\gamma B_{1L} + B_{1R} \times B_{2L}^*])$

$\text{norm}(T_1 \times T_2) = (T_1 \times T_2) \times (T_1 \times T_2)^*$
$= ([B_{1L} \times B_{2L} - \gamma B_{1R}] \times [B_{1L} \times B_{2L} - \gamma B_{1R}]^* + [\gamma B_{1L} + B_{1R} \times B_{2L}^*]^* \times [\gamma B_{1L} + B_{1R} \times B_{2L}^*],$
$-[\gamma B_{1L} + B_{1R} \times B_{2L}^*] \times [B_{1L} \times B_{2L} - \gamma B_{1R}] + [\gamma B_{1L} + B_{1R} \times B_{2L}^*] \times [B_{1L} \times B_{2L} - \gamma B_{1R}])$

$= (\text{norm}(B_{1L} \times B_{2L}) + \text{norm}(B_{1R} \times B_{2L}^*) + \gamma^2 \text{norm}(B_{1R}) + \gamma^2 \text{norm}(B_{1L})$
$-\gamma(B_{1L} \times B_{2L}) \times B_{1R}^* - \gamma B_{1R} \times (B_{1L} \times B_{2L})^* + \gamma B_{1L}^* \times (B_{1R} \times B_{2L}^*) + \gamma(B_{1R} \times B_{2L}^*)^* \times B_{1L}, \mathbf{0})$

To proceed as before we need to show that the norm property $\text{norm}(xy) = \text{norm}(x)\text{norm}(y)$ holds for the bisedenions when one of them is constrained to be in the form of the bisedenion propagator, e.g., does $\text{norm}(B_{1L} \times B_{2L}) = \text{norm}(B_{1L}) \times \text{norm}(B_{2L})$ where B_{2L} is in the form of the bisedenion propagator?

$\text{norm}(B_{1L} \times B_{2L}) = (B_{1L} \times B_{2L}) \times (B_{1L} \times B_{2L})^*$
$= ([S_{1LL} \times S_{2LL} - \beta S_{1LR}] \times [S_{1LL} \times S_{2LL} - \beta S_{1LR}]^* +$
$[\beta S_{1LL} + S_{1LR} \times S_{2LL}^*]^* \times [\beta S_{1LL} + S_{1LR} \times S_{2LL}^*],$
$-[\beta S_{1LL} + S_{1LR} \times S_{2LL}^*] \times [S_{1LL} \times S_{2LL} - \beta S_{1LR}] +$
$[\beta S_{1LL} + S_{1LR} \times S_{2LL}^*] \times [S_{1LL} \times S_{2LL} - \beta S_{1LR}])$

$$= (\; \text{norm}(S_{1LL} \times S_{2LL}) + \text{norm}(S_{1LR} \times S_{2LL}*) + \beta^2\, \text{norm}(S_{1LR}) + \beta^2\, \text{norm}(S_{1LL}) -$$
$$\beta(S_{1LL} \times S_{2LL}) \times S_{1LR}* - \beta S_{1LR} \times (S_{1LL} \times S_{2LL})* +$$
$$\beta S_{1LL}* \times (S_{1LR} \times S_{2LL}*) + \beta(S_{1LR} \times S_{2LL}*)* \times S_{1LL}, \quad \mathbf{0})$$

Now that we've reduced to this level we see there is a problem. In the prior reduction we arrived at the variables being octonions at this stage, for which the norm property and braid laws of the octionoic composition algebra allowed $\text{norm}(O_{1LL} \times O_{2LL}) = \text{norm}(O_{1LL})\text{norm}(O_{2LL})$ and showed the non-norm terms were zero by manipulations using the braid laws that hold for the composition algebras. Now that we've moved to the next higher Cayley algebra's in the derivation, and in our extension construction, we now are asking the sedenions to act as a composition algebra to proceed (on an unrestricted part of the Sedenion algebra). The construction fails. Thus, the extension process does not extend past the Bisedenions, it basically requires the Cayley algebra at two Cayley levels lower to still be a composition algebra. It is still possible to extend to the bisedenions because at two levels lower you still have the octonions, which are a composition algebra as needed. Computationally we see a failure to propagate the bi-trigintaduonions so this is consistent.

2.6 Computational Validation: Code and Results
The key software solution to discover/verify the results computationally is a the recursive Cayley definition for multiplication, which avoids use of lookup tables and avoids commutation and associativity issues encountered at higher order. It is shown next. The cayley subroutine takes the references to any pair of Cayley numbers (represented in list form, so represented as simple arrays), and multiplies those Cayley numbers and returns the Cayley number answer (in list form, thus an array). The main usage was with randomly generated unit norm Cayley numbers that were multiplied (from right) against a "running product". Tests on unit norm hold for millions of running product evaluations in cases where there the unit norm propagations are validated, so, like the perfectly meshed gears of a machine, or the perfectly 'braided' threads of a very long string.

The bignum module was used with 50 decimal places of precision in most experiments, with some experiments at 100 decimal places of precision in further validation testing. Using bignum allows much higher precision handling (needed for the iterative processes of repeated multiplicative updates). The use of bignum, however, entails number representation/storage via strings and is vastly slower than normal arithmetic operations. Furthermore, modern GPU enhancements are not possible with the string handling intermediaries, so the resultant computational threads are CPU intensive and slow.

---------------------------------- cayley_multiplication.pl ----------------------------------

```perl
sub cayley {
    my ($ref1,$ref2)=@_;
    my @input1=@{$ref1};
    my @input2=@{$ref2};
    my $order1=scalar(@input1);
    my $order2=scalar(@input2);
    my @output;
    if ($order1 != $order2) {die;}
    if ($order1 == 1) {
        $output[0]=$input1[0]*$input2[0];
    }
    else{
        my @A=@input1[0..$order1/2-1];
        my @B=@input1[$order1/2..$order1-1];
        my @C=@input2[0..$order1/2-1];
        my @D=@input2[$order1/2..$order1-1];
        my @conjD=conj(\@D);
        my @conjC=conj(\@C);
        my @cay1 = cayley(\@A,\@C);
        my @cay2 = cayley(\@conjD,\@B);
        my @cay3 = cayley(\@D,\@A);
        my @cay4 = cayley(\@B,\@conjC);
        my @left;
        my @right;
        my $length = scalar(@cay1);
        my $index;
        for $index (0..$length-1) {
            $left[$index] = $cay1[$index] - $cay2[$index];
            $right[$index] = $cay3[$index] + $cay4[$index];
        }
        @output=(@left,@right);
    }
    return @output;
}
```

---------------------------------- cayley_multiplication.pl ----------------------------------

2.6.1 Unit-norm multiplicative 'step' generation method

The randomly generated propagation step is for a unit norm that has a randomly generated small perturbation. Consider the eight element octonion denoted: $\{\Delta x_0, \delta x_1, \delta x_2, ... \delta x_7\}$, where the real component is $\Delta x_0 \approx 1$:

$$(\Delta x_0)^2 = 1 - \Sigma (\delta x_i)^2 ,$$

And where each δx_i is generated by a randomly generated number uniformly distributed on the interval (-0.5 .. 0.5), with an additional perturbation-factor 'δ', e.g., the max magnitude imaginary perturbation from pure real ($\Delta x_0 = 1$), measured with L_1 norm, is simply 7 times $\delta/2$ (for seven imaginary components).

For the octonions unrestricted unit norm propagation is possible, i.e., all of the components can be independently generated and then normalized to have L_2 norm $=1$. So, the restriction to $\mathbf{\Delta x_0}{\approx}1$ isn't needed. The same is true on the (left) chiral extension spaces:

Propagating chiral left sedenion: $\{\Delta x_0, \delta x_1, \delta x_2,...\delta x_7, \delta x_8\}$, with $\delta x_9=0, ..., \delta x_{15}=0,$

and the

Propagating chiral left bi-sedenion: $\{\Delta x_0, \delta x_1, \delta x_2,...\delta x_7, \delta x_8, \delta x_{16}\}$, with $\delta x_9=0, ..., \delta x_{15}=0, \delta x_{17}=0, ..., \delta x_{31}=0.$

Consider now the propagating chiral left bi-sedenion with the a small non-propagating component (here δx_9 nonzero is chosen), and now let's return to formally requiring $\mathbf{\Delta x_0}{\approx}1$ on propagation steps:

Propagating small perturbation chiral left bi-sedenion: $\{\Delta x_0, \delta x_1, \delta x_2,...\delta x_7, \delta x_8, \mathbf{\delta x_9}, \delta x_{16}\}$, with $\delta x_{10}=0, ..., \delta x_{15}=0, \delta x_{17}=0, ..., \delta x_{31}=0.$

For the small perturbation steps that are randomly generated, there are now ten imaginary components, so in what follows, the maximum magnitude of the imaginary components, measured with L_1 norm, denoted $\mathbf{\Delta}$, is $\mathbf{\Delta}=10\delta/2=5\delta.$

2.6.2 Computational Results

Norm, N, decay (from 1) with propagations on bi-sedenions with δx_9 nonzero have been examined on repeated dataruns consisting of 1,000,000 multiplicative (small path step) iterations each. The decay from '1' is significant in the initial studies, so only the exponent of the norm is shown initially:

δ	N (at 1M iter.)	Index of 1st rczc	#rczc (at 1M iter.)
0.5	e-36	14	75291 (at 0.5M)
0.5	e-43	19	73982 (at 0.5M)
0.5	e-38	9	75053 (at 0.5M)
0.5	e-35	11	75369 (at 0.5M)
0.5	e-35	36	74769 (at 0.5M)

Table 1. Repeated runs showing a range of different norm, N, decay with propagations on bi-sedenions with δx_9 nonzero ("rczc" is used for 'real component zero-crossing').

There appears to be randomness on the choice of slope (fall-off), but once propagating many iterations, the choice that is selected appears to be kept (i.e., is an emergent, stable over many iterations, structure, in the fall-off behavior). A similar range of randomness in fall-off curves appears for the other parameters if unfrozen instead – Table 2 shows this where δx_{10} is nonzero instead of δx_9.

δ	N (at 1M iter.)	Index of 1st rczc	#rczc (at 1M iter.)
0.5	e-44	31	74878 (at 0.5M)
0.5	e-34	13	75088 (at 0.5M)
0.5	e-44	23	74527 (at 0.5M)
0.5	e-45	9	74588 (at 0.5M)
0.5	e-39	9	75108 (at 0.5M)

Table 2. Using δx_{10} nonzero instead of δx_9, again have repeated runs showing a range of different norm, N, decays with propagations on bi-sedenions ("rczc" is used for 'real component zero-crossing').

For Table 3, 4, & 5, we consider smaller perturbations, with delta=0.01:

δ	N (at 1M iter.)	Index of 1st rczc	#rczc (at 1M iter.)
0.1	0.23128	426	27737
0.1	0.86074	300	28718
0.1	0.79330	872	28490
0.1	0.85927	488	29144
0.1	0.39766	527	29644

Table 3. Repeating with the perturbation reduced to 0.1 with propagations on bi-sedenions with δx_9 nonzero ("rczc" is used for 'real component zero-crossing').

δ	N (at 1M iter.)	Index of 1st rczc	#rczc (at 1M iter.)
0.1	0.44672	619	28759
0.1	0.81644	416	28652
0.1	0.46067	278	29589
0.1	0.95643	933	28382
0.1	0.79521	281	28569

Table 4. Repeating with the perturbation reduced to 0.1 with propagations on bi-sedenions with δx_{10} nonzero ("rczc" is used for 'real component zero-crossing').

δ	N (at 1M iter.)	Index of 1ˢᵗ rczc	#rczc (at 1M iter.)
0.1	1	358	26807
0.1	1	555	27241
0.1	1	580	27534
0.1	1	395	26625
0.1	1	618	26990

Table 5. Repeating with the perturbation reduced to 0.1 and considering propagations with all non-propagating components zero, e.g., back to a test of unit-norm propagation on the chiral bisedenion subspace.

Note how even in the norm preserving propagation, the context of 0.1 non real-component mixing 'perturbation' still leads to a zero-crossing for the real component (that starts at 1) similar to the non-propagating perturbations considered. Evidently, the mixing seen under multiplication during each propagation step can be very significant even at perturbation of 0.1. So let's consider 0.01, with results shown in Tables 6, 7, & 8.

δ	N (at 100K iter.)	Index of 1ˢᵗ rczc	#rczc (at 100K iter.)
0.01	1.01143	48027	156
0.01	0.99173	0	0
0.01	1.021529	82489	24
0.01	1.0009355	31342	316
0.01	1.0013	53429	50

Table 6. Repeating with the perturbation reduced to 0.01 with propagations on bi-sedenions with δx_9 nonzero.

Note that the zero counts on the second datarun aren't particularly significant since the third run had its first zero-crossing at 82489, This simply presents the likely possibility that the first zero crossing in the second run simply did not occur in the first 100K iterations under observation.

δ	N (at 100K iter.)	Index of 1st rczc	#rczc (at 100K iter.)
0.01	1.00305	49631	325
0.01	1.00272	83242	63
0.01	1.001873	62510	166
0.01	1.000101	43666	178
0.01	0.9956	53451	219

Table 7. Repeating with the perturbation reduced to 0.01 with propagations on bi-sedenions with δx_{10} nonzero.

δ	N (at 100K iter.)	Index of 1st rczc	#rczc (at 100K iter.)
0.01	1	32901	139
0.01	1	43551	302
0.01	1	81880	56
0.01	1	34585	542
0.01	1	37826	111

Table 8. Repeating with the perturbation reduced to 0.01 and considering propagations with all non-propagating components zero, e.g., back to a test of unit-norm propagation on the chiral bisedenion subspace.

If we repeat with δ=0.001, we get approximate unit norm preservation (e.g., a stable oscillation about N=1 appears to occur), where mixing is never so significant that there occurs a zero-crossing in the real component (see Table 9 for summary).

δ	N (at 1M iter.)	Index of 1st rczc	#rczc (at 1M iter.)
0.5	e-36	14	75291 (at 0.5M)
0.1	0.23128	426	27737
0.01	1.01143	48027	156 (at 0.1M)
0.001	1.0000149	n/a	0

Table 9. Summary of norm propagation results with delta at different scales.

If we wish to find the maximal δ where mixing is never so significant that even a single zero-crossing occurs in the real component we have the results shown in Table 10 (only the δx_9 nonzero perturbation propagation case is shown).

δ	N (at 1M iter.)	Index of 1st rczc	#rczc (at 1M iter.)
0.007	1.0010	210774	1777
0.006	1.0040	198967	1743
0.005	0.99998	175311	1451
0.004	0.99855	136624	828
0.003	0.999745	449593	229
0.002	0.999965	457410	909
0.0015	0.9990772	868253	224
0.001475	0.999990	972837	27
0.00146	0.9997605 (2M)	1884455	133 (at 2M iter.)
0.0014595	1.0000569 (2M)	1886191	219 (at 2M iter.)
0.0014585			
0.0014575	1.000249 (2M)	n/a	0 (at 2M iter.)
0.001455	1.000011 (2M)	n/a	0 (at 2M iter.)
0.00145	0.999883	n/a	0
0.0014	0.9999989	n/a	0
0.0013	1.0000109	n/a	0
0.00125	1.00016825	n/a	0

Table 10. Summary of norm propagation results from a search for the max delta which permits approximate norm =1 propagation, with no real-component decay to non-positive allowed.

The maximal perturbation parameter δ allowing approximate norm=1 propagation, with no real-component decay to non-positive allowed (i.e., no zero-crossing), is shown for δ=0.0014575 on an iteration window of 2,000,000. From Table. 10, we see that the max delta with #rczc=0 lies somewhere between 0.0014575 and 0.0014595, and estimating this to be the midpoint, we have the estimated max delta to be: 0.0014585 (as shown in the Table).

As mentioned in the Methods, for the small perturbation steps that are randomly generated there are ten imaginary components. Evaluating the magnitude of the perturbation in terms of the relation between real component (approximately 1) and the maximum magnitude of the imaginary components, measured with L_1 norm, denoted **Δ**, we have **Δ**=10δ/2=5δ. Thus, the maximum perturbation allowed for unitary propagation is estimated to be **Δ=.0072925,** which is the fine structure constant, where: **1/Δ=137=1/α.**

In this computational exploration we restrict the propagation mechanism to be multiplication (on the right, say) by a unit norm element that is a small perturbation from the unit element. In this analysis we see a precipitous phase transition in propagation behavior when the perturbation becomes sufficiently small, and this is more than the incremental change in perturbation would suggest. The maximal perturbation magnitude allowed for a propagating path construction mechanism has been experimentally determined to be none other than alpha, the fine structure constant, hitherto only determined experimentally.

Now consider a sum of such emanations that results in a propagator with stationary phase (when compared to similar propagation histories), along with the familiar classical, semiclassical, and quantum behavior. As with the standard sum on paths construction, the phases of paths without stationary phase cancel and are eliminated from consideration. Note that the trigintaduonion emanation eliminates propagation not just by the non-perturbative non-unit-norm decay trajectories within the trigintaduonions, but the higher Cayley non-decaying (divergent norm) trajectories. (Not shown are propagation efforts for Cayley algebras higher than Trigintaduonion (or Bi-Sedenion). At the next higher, 64-element, algebra, consideration of perturbation with a small δx_{32} component, rather than nonzero δx_9 component, yield norm evaluations that diverge very rapidly.) ***In effect, the fine structure constant is selected to allow for maximal perturbative propagation in the 32D Trigintaduonion space.*** So the unified propagator theory appears to suggest maximal unitary propagation in the context of unit norm multiplication by perturbed unit elements with imaginary component having magnitude less than alpha. If 'reality' wanted to propagate 'information' within a single algebra construct, and allow for maximal information transmission, that object would evidently be an element with 'alpha'= maximum perturbation.

Propagation with the max-perturbation alpha step then leads to emanation trajectories that can be examined insofar as their as far as real-component zero-crossing times (and emergent fall-off from unity trajectories if max-perturbation exceeds alpha). Further analysis in this context is done in Ch. 10 on the computational evidence that the Emanation process is Martingale.

Chapter 3. Chiral Trigintaduonion Emanation

3.1 Chiral T-emanation has 78 generators of change and 137 independent octonion terms

We begin with constructing the theoretical expression for a general element of the trigintaduonion algebra after two chiral trigintaduonion multiplicative propagation steps. A simple analysis of the number of terms in this expression, when reduced to three-element algebraic 'braid-level', results in a count on algebraic braids of 137, plus a little extra (e.g. some lagniappe for the best 'cooking') of a contribution towards a 138th braid when the "noise analysis" is done.

3.1.1 Trigintaduonion Emanation and the Critical Parameters 78, 137 and α [1,6,15,16]

Consider a general Norm=1 (32D) Trigintaduonion (Bi-Sedenion): (A,B), where A and B are sedenions (16D). Then have (A,B) = ((a,b), (c,d)), where {a,b,c,d} are octonions. Slightly different than a propagator, we have an 'emanator' with the following notation and properties, where the emanator describes a 10D multiplicative step. The emanator is a chiral bi-sedenion: a trigintaduonion whose first sedenion half is itself a chiral bi-octonion, and the second sedenion half is a pure real (as is the second octonion half): $(\tilde{A},\beta)$, $\tilde{A} = (\tilde{a},\alpha)$, where the norm is 1, α is a real octonion, and β is a real sedenion. Thus:

Emanator: $(\tilde{A},\beta) = ((\tilde{a},\alpha), \beta)$.
Note: $\tilde{A}^* = (\tilde{a}^*,-\alpha)$.

Let's set up a description of the Universal 'Emanation' along a 'chiral path' resulting from a few emanation steps. To begin, suppose we have already arrived at, or received, a unit norm trigintaduonion (32D) state 'T', and suppose our emanations are the result of right multiplication with a chiral trigintaduonion (bi-sedenion) 'step', and suppose we consider one such path after just a few steps. Here's the notation to begin:

T = (A,B), a unit norm trigintaduonion.
τ = $(\tilde{A},\beta)$ = ($(\tilde{a},\alpha)$, β), the 'emanator' above (so named to distinguish from a 'propagator').

Universal Emanation from T on single path with three steps: $(((\mathbf{T} \bullet \tau_1) \bullet \tau_2) \bullet \tau_3) \ldots$

Consider the first emanation step:

$\mathbf{T} \bullet \tau_1 = (A,B) \bullet (\tilde{A},\beta) = ([A\bullet\tilde{A}-\beta^*\bullet B] , [B\bullet\tilde{A}^*+\beta\bullet A])$. (Standard Cayley algebra multiplication rules.)

$A\bullet\tilde{A} = (a,b) \bullet (\tilde{a},\alpha) = ([a\bullet\tilde{a}-\alpha^*\bullet b] , [b\bullet\tilde{a}^*+\alpha\bullet a])$

$B\bullet\tilde{A}^* = (c,d) \bullet (\tilde{a}^*,-\alpha) = ([c\bullet\tilde{a}^*+\alpha^*\bullet d] , [d\bullet\tilde{a}-\alpha\bullet c])$

Thus,

$\mathbf{T} \bullet \tau_1 = (A,B) \bullet (\tilde{A},\beta) = ([(a\bullet\tilde{a}-\alpha^*\bullet b-\beta c) , (b\bullet\tilde{a}^*+\alpha\bullet a-\beta d)] , [(c\bullet\tilde{a}^*+\alpha^*\bullet d+\beta a) , (d\bullet\tilde{a}-\alpha\bullet c+\beta b)])$.

At the lowest octonion level, that covers the pure real trigintaduonion, we have:

$(a\bullet\tilde{a}-\alpha^*\bullet b-\beta c) \rightarrow 8\times8 + 8 + 8 - 2 = 64+14 = 78$ independent octonion terms (78 independent generators of motion). The -2 comes from the unit norm constraints on T and τ.

Now consider the second propagation step:

$(\mathbf{T} \bullet \tau_1) \bullet \tau_2 = ([(a\bullet\tilde{a}-\alpha^*\bullet b-\beta c) , (b\bullet\tilde{a}^*+\alpha\bullet a-\beta d)] , [(c\bullet\tilde{a}^*+\alpha^*\bullet d+\beta a) , (d\bullet\tilde{a}-\alpha\bullet c+\beta b)]) \bullet (\tilde{A},\beta)$,

where $\tau_2 = (\tilde{A}',\beta') = ((\tilde{a}',\alpha'), \beta')$.

Let $(\mathbf{T} \bullet \tau_1) \bullet \tau_2 = ([Z_{11},Z_{12}] , [Z_{21}, Z_{22}])$.

$Z_{11} = (a\bullet\tilde{a}-b\alpha-c\beta)\bullet\tilde{a}' - (b\bullet\tilde{a}^*+\alpha a-\beta d) \alpha' - (c\bullet\tilde{a}^*+d\alpha+a\beta)\beta'$.

In Z_{11} we can replace the octonions with their unit component forms:

$$a = a_1 e_1 + a_2 e_2 + \ldots + a_8 e_8 ,$$

where $\{e_1, e_2, \ldots, e_8\}$ are the unit octonions (one real, seven imaginary), while 'α'$=\alpha e_9$ and 'β'$=\beta e_{17}$, originally, but in expressions, are reduced to just their real part. All expressions, thus, involve 10 components: $\{e_1, e_2, \ldots, e_8, e_9, e_{17}\}$, and as the equations for Z_{11} shows, grouped in factors of three (three-element octonionic 'braids'). We don't have associativity but we do have alternativity and the braid rules on three-element octonionic products that allows their regrouping. Applying these rules to have only ordered $e_i\bullet e_j\bullet e_k$ products in a simplified expression, we will then have $10\times9\times8/3! = 120$ independent terms when the products involve different components. We have 8 independent terms when the first product are on

34

the same component (equals 1), have 8 independent terms when the second product involves the same component, and have 1 independent term when the three-way product equals 1 (further details on this and the properties of the exponentiation map on hypercomplex numbers is given in the next section. There are, thus, 137 independent terms in Z_{11}, where each term has norm less than unity (since each octonionic component has norm less than one and the norm of a product of octonions is the product of their norms). The terms involving products with the same component, or with the components three-way product equal unity, correspond to the 'telescoping terms' in what follows.

When $T=((a,b),(c,d)) \rightarrow ((T \bullet \tau_1) \bullet \tau_2)=((Z_{11},Z_{12}),(Z_{21}, Z_{22}))$. we have $a \rightarrow Z_{11}$ and the terms involving 'a' in Z_{11} are referred to as 'telescoping' due to their simple math properties with further emanation steps. In particular, the terms involving 'a' are:

Z_{11}[a terms]$= a \bullet \tilde{a} \bullet \tilde{a}' - a\alpha\alpha' - a\beta\beta'$.

We can see that the original 'a' information is passed along three (telescoping) channels, one involving repeated full octonionic factors $\tilde{a}$, one involving repeated real-octonion α factors, and one involving repeated real-octonion β factors:

(1) a $\rightarrow$ (a$\bullet\tilde{a}$)$\bullet\tilde{a}'$, if this product is continued indefinitely, then we have *the random product of a collection of octonions*, all of which have norm less than one (although their norms can be quite close to one). If their norms were perfectly equal to one, then the addition of their random 'phases' would tend to cancel to zero, giving only a real octonionic component (same argument for phase cancelation on S1 as on S7 or S15). What results is a 'mostly' real octonion, having some imaginary part. A more precise, and lengthy, derivation is given in the next section.

(2) a $\rightarrow$ a$\alpha\alpha'$, if this product is continued indefinitely, 'telescoped' with repeated α products, we see that the original 8 independent terms arising from 'a' are passed forward with an overall real octonion product, giving rise to 8 independent terms.

(3) a $\rightarrow$ a$\beta\beta'$, as with (2), we have 8 independent terms.

From the above, we see an alternative accounting of the extra 17 independent terms to go with the 120 for a total of 137 independent terms in the propagation of the octonionic sectors of the universal emanation. A benefit of the telescoping analysis is it clarifies how in (1) an imaginary component may arise, and in perturbation expansions it will then be natural to refer to an overall imaginary component.

There are 137 terms in the dually chiral 'emanation', each with norm bounded by unity, with total bi-sedenion norm equal to unity. In the analysis that led to the computational discovery of α [6], an imaginary (non 10D) component was added of growing magnitude until unit-norm propagation failed. In essence, a maximum perturbation, from propagation strictly in the 10D subspace of the 32D trigintaduonions, was sought.

We identify maximal perturbation by doing an independent term analysis, and by adding a maximum perturbation term that implicitly identifies a definition of maximum antiphase. From this definition of maximum antiphase, there results the parameter π.

3.1.2 Summary with detailed 137th count and the Exponential Map on Hypercomplex Numbers

The derivation below follows [22,37], but with a more succinct accounting of the independent terms.

Consider a general norm=1 bisedenion in list notation: (A,B), where A and B are sedenions. Consider a propagator bisedenion (C,β), $C = (c,\alpha)$, where c is an octonion and α is shorthand for the real octonion $(\alpha,0,0,0,0,0,0,0)$, where α is a real number, and β is shorthand for the real sedenion $(\beta,0,0,0,0,0,0,0,0,0,0,0,0,0,0,0)$, where β is a real number. Using A=(a,b), B=(u,v), and the multiplication rule from Sec. 2, we have:
$(A,B)(C,\beta) = ([AC-\beta^*B], [BC^*+\beta A])$, where
$AC = (a,b)(c,\alpha) = ([ac-\alpha^*b],[bc^*+\alpha a]); BC^*=(u,v)(c^*,-\alpha) =$
$([uc^*+\alpha^*v],[vc^*-\alpha u])$.
Thus, we have:
$(A,B)(C,\beta) = ([(ac-\alpha^*b , bc^*+\alpha a)-\beta^*(u,v)] , [(uc^*+\alpha^*v , vc-\alpha u)+\beta(a,b)])$,
so,
$(A,B)(C,\beta) = ([ac-\alpha^*b-\beta^*u , bc^*+\alpha a-\beta^*v] , [uc^*+\alpha^*v+\beta a , vc-\alpha u+\beta b])$.
Now consider another propagator bisedenion (C',β'), $C' = (c',\alpha')$, and form the product corresponding to the next multiplicative step:
$((A,B)(C,\beta)) (C',\beta') = ([(ac)c' - \alpha^*bc' - \beta^*uc' - \alpha'^*(bc^*+\alpha a-\beta^*v) , ...] ,$
$[... , ...])$, where only the first expression at octonionic level (
$T=(O_1,O_2,O_3,O_4)$) is shown:

36

$$O_1 = (ac)c' - \alpha^*bc' - \beta^*uc' - \alpha'^*(bc^* + \alpha a - \beta^* v).$$

At octonionic-level there are 10x9x8/3x2=120 independent terms for 8 octonionic components (labeled a, b, c) plus a separate octonion component (α) and one sedenion component (β), e.g., have 10 choose 3. Also have telescoping terms with repeated real octonion factors, such as with the $a\alpha\alpha'^*$ term (think $a\alpha(\alpha'^*)^n$), which gives an additional 8 independent terms. Also have telescoping terms with alternating real octonion factors and real sedenion factors, such as with the $v\beta^*\alpha'^*$ term (think $v(\beta^*\alpha'^*)^n$), which gives another 8 independent terms. There is one other 'telescoping' term due to repeated octonion right products seen in $(ac)c'$ (now think $((ac)c')c'.....c')$). The change in this term corresponds to an element of the automorphism group on octonions, G2, and as such provides one last independent term, for a total of 137 independent terms at octonion level.

All of the octonion products involve octonions with norms at most unity, and by the normed division algebra rules on octonions, their norm is simply the norm of the individual octonions multiplied together, all of which are bounded by unity, thus their product is bounded by unity. The overall bound for the expression, each individual term being bounded by unity, is therefore simply the counting on the independent terms.

The maximum magnitude of each component of the octonion in the product term is given with a 'channel multiplier' of 137. Also, in seeking the maximum information propagation we require that the real chiral component never cross zero (e.g., stay in its connected $\{\alpha, \beta\}$ quadrant), thus the strictest condition on evaluating evolution might be intuited to be when the imaginary components combine to have real component contribution that is antiphase, e.g., the total imaginary angle is π. The choice of antiphase will used in what follows and will be justified when "C ×" allows the antiphase to be understood in the context of the Universal Mandelbrot set [38] position on the negative real axis that gives the maximal magnitude of displacement from the origin: C_∞. We limit the maximum perturbation allowable by the antiphase worst case. At octonionic-level there is thus the channel multiplier: $137 + i\pi$.

For what follows, it helps to recall some important properties of the exponential, particularly its well-defined properties with hypercomplex numbers [22]. **Important map relations:**

(1) exponential map on Im(T) gives unit norm object: $\exp(\text{Im}(T)\theta) = \cos\theta + \text{Im}(T)\sin\theta$.

(2) exponential map on iT gives $C \times T$:
$\exp(iT) = \exp(i\text{Re}(T)) \times \exp(i\text{Im}(T)) = (\cos\theta + i\text{Re}(T)\sin\theta) \times (\cos\varphi + i\text{Im}(T)\sin\varphi) = C \times T$

Use (1) to focus on fluctuations in imaginary parameters free of normalization concerns.

Use (2) to get complex structure $C \times$ (object). Note that exponentiation into phase terms is precisely what occurs in the path integral propagator formalism, and will occur here as well for the emanator formalism, thus the "$C \times$" complex factor. When drawn upon in the emanator formalism, this method of achieving additional "$C \times$" complex structure will be forced by the zero-divisor handling (that will give rise to point-like matter with very small phase coupling, thus a highly oscillatory integral, and ties over to foundational aspects of the path integral formalism).

3.2 Chiral T-emanation has max perturbation alpha
3.2.1 The {α, π} relation
In the methods we saw that there are 137 independent tri-octonionic braid propagations contributing to the overall chiral trigintaduonion propagation (further details in [6]), in each of its octonion subparts, along with an independent imaginary component (in those sub-parts with 137 terms). At the component level of the base trigintaduonion, we similarly have 137 independent (real) terms, each with maximum one, thus an evaluation of the maximum at the component level involves a simple counting on the (unit max) independent terms. Aside from an overall scale factor, the maximum magnitude at component level involves a real part of magnitude 137 and an imaginary part. We hypothesize that the imaginary part has magnitude π in relation to the 137, for maximum antiphase when viewed as a phase angle (to be justified in the next paragraph), and we thereby arrive at component-level having an overall maximum perturbation given by $137+i\pi$, i.e. an overall perturbation magnitude of the injected perturbation amount δ and the multiplier $137/\cos(\pi/137)$.

At trigintaduonion level, we see that the overall maximum perturbation is given by the individual component-level perturbation amounts in the chiral emanation and their possible convergence into a maximum

38

magnitude factor of 137/cos(π/137), for maximum perturbation amount δ x 137/cos(π/137) (see Fig. 1). Thus, at the level of the independent terms (137) in each of the chiral trigintaduonion independent components (29), each such term has a maximum perturbation contribution with magnitude δ x 137/cos(π/137), each with phase angle $\theta=(\pi/29\times137)$ to have equipartitioning of phase among the 29x137 independent terms (see Fig. 2).

According to the Kato-Rellich Theorem (described in the next section), the maximum perturbation is such that the magnitude of the total perturbation is ≤1 (as described in the Methods). Before we can do this step, however, we must rescale such that component-level imaginary component equals component level phase (thereby introducing a factor $\theta/\sin\theta$, see Fig. 3). This is a result of the normalization step in the achiral emanator, the existence of which is related to the hypothesis that component sums are made interchangeable with angle sums (an effect of the exponential map on hypercomplex numbers described in a later section). Here the result is we arrive back at a component-level sum of all of the imaginary parts totaling π, which was the initial hypothesis, and we have for maximum perturbation

$$\alpha_{max} = (1/137)(\cos\beta/\cos\theta)(\sin\theta/\theta),$$

where $\beta = (\pi/137)$ and $\theta = \pi/(137 \times 29)$.

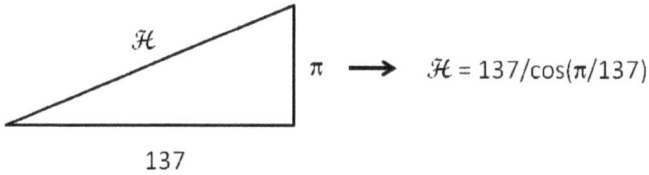

Fig. 1 The magnitude relation at Trigintaduonion-level.

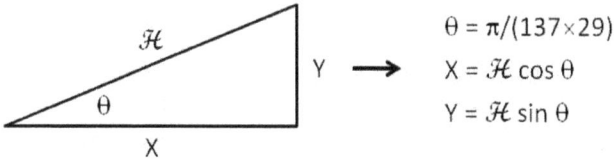

Fig. 2. The magnitude-angle relation at independent terms level, given 29 independent dimensions and 137 independent terms in each.

39

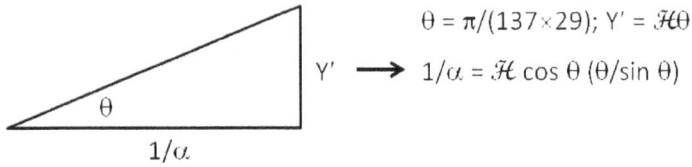

$\theta = \pi/(137 \times 29)$; $Y' = \mathcal{H}\theta$

$Y' \longrightarrow 1/\alpha = \mathcal{H}\cos\theta\,(\theta/\sin\theta)$

$1/\alpha$

Fig. 3. Emanator definition gives imaginary component sums are made interchangeable with angle sums (given sum with normalization in definition). This can be described in terms of analyticity in general or Euclideanizability.

To recap, First, consider a trigintaduonion element of propagation that results from multiple achiral emanation steps, for which it's octonion subsectors will have 137 independent terms (resulting from tri-octonion products) with perturbation (or noise) magnitude having a factor of $\mathcal{H}=|137+i\pi|$ (see Fig. 1), where the unit norm upper bound on the tri-octonion products gives the 137 and the "maximal antiphase" π phase amount is justified and made self-consistent, at the next step.

Second, now consider the maximal noise element at the level of each 137 independent term in each of the 29 (free) dimensions in each of the chiral product terms in $\mathbf{T} \bullet T^{(k)}_{chiral}$ in the Emanator (in essence, interpret the multiplication as projecting the other way, $\mathbf{T}$ onto the chiral basis specified by $T^{(k)}_{chiral}$). Again, we postulate that the total imaginary amount will be at maximal antiphase, or such that the amount of phase for each of the 137x29 independent terms is $\theta=(\pi/29\text{x}137)$, indicating the general relation shown in Fig. 3.

Now consider the magnitude rescaled in Fig. 3 such that the hypotenuse is 1 (unit norm), it is then clear that the maximum allowed perturbation, $1/\alpha$, satisfies $((1/\alpha)/\mathcal{H})=(\theta/\tan\theta)$ (see Fig. 4). Note the distinctive arrangement that the maximal noise, or perturbation, hypothesis reveals in Fig. 4, where the phase angle and imaginary component value are equal (already suggested by both component-level sum in the Emanator, and phase-angle sums from the chiral product terms, must total maximum antiphase π). This is simply an artifact of an exp map relation, and will be discussed further in a later section.

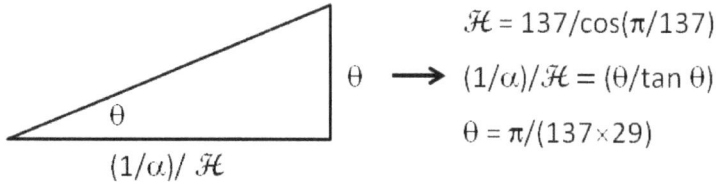

The figure shows a right triangle with base labeled $(1/\alpha)/\mathcal{H}$ and an angle θ at the left vertex (also marked inside). To the right of the triangle:

$$\mathcal{H} = 137/\cos(\pi/137)$$
$$\theta \longrightarrow (1/\alpha)/\mathcal{H} = (\theta/\tan\theta)$$
$$\theta = \pi/(137 \times 29)$$

Fig. 4 Unit norm case. $((1/\alpha)/\mathcal{H})=(\theta/\tan\theta)$. Shows phase angle = imaginary component magnitude. .

Thus, maximal noise, or perturbation transmission occurs when the noise phase angle equals the noise imaginary component, when noise scaled to total magnitude 1. This resulted in part by assumptions going into the construction of an achiral emanator from summing over chiral emanations (multiplications), and would generally result for a variety of such emanators. Further specifics of the emanator construction are required for the next chapter, however, so this emanator dependency will be developed further there. Before moving on, however, it appears that the only constraint on emanators would be that they generate, through inclusion of unbiasing sums on chiral multiplications, that the noise phase angle equals the noise imaginary component relation. I'll refer to this relation as the proto-Euclideanizability, or proto-analyticity, property of the Emanator. If we start with the hypothesis that the Emanator will induce a proto-Euclideanizability relation, this allows us to start directly at Fig. 4 in evaluating the maximum perturbation allowed, and we get fundamental Euclideanizability as a side effect. Regardless of starting hypotheses, the end result for the maximum perturbation magnitude is

$\alpha = (1/137)(\cos\beta/\cos\theta)(\sin\theta/\theta)$, where $\beta = (\pi/137)$ and $\theta = \pi/(137 \times 29)$.

Thus, $1/\alpha \cong 137.0359998$, where the last digit is uncertain given the precision used (this relation originally appeared in [39] but without explanation in terms of trigintaduonions).

Since α is a fundamental parameter that emerges for a maximal propagation, and we find here another relation on α that ties it to the maximal antiphase amount 'π', taken in reverse, this offers an alternate origin for the fundamental parameter π from mathematics. Although the

41

idealizations of planar geometry can be used to derive π (or modern variants from complex analysis involving the complex plane) it is interesting that we have here an origin of π via what leads to maximal anti-phase when computing α_{max}, where $\alpha=\alpha_{max}$ is selected for maximal information propagation.

3.2.2 The {α, π} relation with explicit use of Exp Map properties

The maximum perturbation, referred to as maximum noise in what follows, is first evaluated for a chiral emanation where we take a norm=1 $T_{base}=(A,B)$ and take the right product with T_{chiral} in the form $T_{chiral}=(C,\beta)$, with product $(A,B)(C,\beta)$ proven to be unit norm above [1]. In the prior section we saw that there are 137 independent octonion terms at the octonion sub-level of the new unit norm trigintaduonion that results, which leads to 137 independent terms at component level. In order to use the map rules mentioned in the previous section, it is necessary to move from the trigintaduonion, T, space, to the C × T space. This is done in later sections anyway where we consider sums on exp(iT). The exponential function (map) provides a well-defined 'lift' of a hypercomplex (Cayley) algebra from T to C × T. The exponential map also provides a very useful maneuver when working with unit-norm hypercomplex numbers via the generalized deMoivre theorem exp(Im(T))=cos(θ)+Im(T)sin(θ), with the real part recoverable from cos(θ). More details on this follow later but for now, in evaluating the maximum noise allowed we have three structures to adopt: (1) the noise is generalized to be complex (as will be the case for the components themselves once the T→C × T structure is adopted). (2) At component-level, the noise (for maximum noise) is equipartitioned in both real and imaginary parts. (3) Total imaginary noise magnitude is π for maximal antiphase (to be justified later).

(I) Chiral emanation noise: have 137 terms with max unit norm each, for the real part, and for the imaginary part have a "phase angle" β such that $137\beta=\pi$ (here referred to as a phase angle in the sense that the exp(Im(T)) map is being used). The noise magnitude at octonionic-component level is then given by the right triangle with real part = 137 and angle $\beta=\pi/137$, thus maximum chiral emanation noise magnitude is:

$$H = 137/\cos(\pi/137)$$

(II) Achiral emanation noise: now have 29 "free" components, each with 137 independent terms. For maximum achiral emanation we thus have

137 x 29 independent terms that are built from the aforementioned chiral emanation terms (to make achiral). If we equipartition as before, with noise magnitude Hc, we have a "noise triangle" with magnitude (hypotenuse) Hc and with angle $\theta = \pi/137x29$. The imaginary part is then (Hc)sin(θ). As regards the H magnitude separated form (separating out the 'H' factor for now), we have for the imaginary part sin(θ)c. As before, we take maximal noise transmission when all the imaginary parts add to maximal antiphase. Given the equipartitioning assumption, we then simply have the factor 137x29:

$$\sin(\theta)\, c\, (137x29) = \pi \rightarrow c = \theta/\sin\theta.$$

The maximum real noise perturbation that the system can have is then α, where:

$$\alpha^{-1} = \frac{137}{\cos\beta}\cos\theta\,\frac{\theta}{\sin\theta}, \qquad where\ \beta = \frac{\pi}{137}\ and\ \theta = \frac{\pi}{137x29}$$

$$\alpha^{-1} = 137.03599978669910,$$

where the evaluation was done at WolframAlpha to high precision [40] (e.g., higher precision than that reported in earlier work [6]). This matches the experimentally observed value to all 11 decimal places currently known. As of 2002 [41], the measured value of α is:

$$\alpha^{-1} = 137.03599976(50).$$

Note that in quantum field theory the parameters are renormalized at a particular energy scale. Thus choice of energy scale impacts the value of α (as a coupling constant in the classical theory or a perturbation expansion factor in the quantum theory). At 0K we have the extreme low-energy end of the renormalization group (with the largest α value). We are at the 2.7K CMBR, so we have the max α to very high precision. (In studies at high energy scale at LEP, at the energy scale of the Z-boson (91GeV), we get the renormalized value to be [41,42]: $\alpha^{-1}[M_Z] \cong 127.5$. Note that 91GeV is way above the energy scale of the familiar Hagedorn temperature at ~pion mass=150MeV or 1.7x10^12 K) [43], where hadronic matter 'evaporates' into quark matter.)

3.2.3 Kato-Rellich Theorem and Noise Budget Analysis [44]
Definition of "B is A-bounded": Let $A: D(A) \rightarrow H$ be a self-adjoint operator. Let $B: D(B) \rightarrow H$ be symmetric. If $D(A) \subset D(B)$ and $\exists$ b s.t.

$\|Bf\| \le a\|Af\| + b\|f\| \ \forall \ f \epsilon D(A)$, then B is A-bounded with bound a.

Kato-Rellich Theorem: Let $A: D(A) \to H$ be a self-adjoint operator. Let B be A-bounded with bound $a < 1$. Then $A + B: D(A) \to H$ is self-adjoint.

Corollary to Kato-Rellich Theorem: If K-R theorem applies and A is bounded below, then so is $A + B$.

The above corollary is significant in the analysis that follows since the forms of the emanator (sums over 4-suits, or 72deck, or 78deck, especially), before normalization, are here seen to be bounded from below if within the perturbation-limit of the sum-type emanation considered. This means that the normalization step won't fail (divide by zero), furthermore, it indicates no zero divisors when operating within the perturbation limit when considering a single chiral path. Interestingly, the reverse is also indicated, a loss of boundedness just beyond the perturbation limit, including the existence of zero-divisor "land mines".

To apply this to our trigintaduonion analysis, let's "lift" the trigintaduonion T into a formal operator setting as a T-position operator, whose distance operator (from the origin) is the norm(T). Let's denote the Emanation of T by one step by $Em_\delta^{(n)}(T)$, where $\delta = 0$ is the case for no perturbation. We arrive at the form necessary for a self-consistent emanation rule (with well-defined sum from above) when:

$$\left\| Em_\delta^{(n)}(T) \right\| \le n^*\delta \left\| Em_0^{(n)}(T) \right\| + b\|f\|,$$

where choosing b=1 for simplicity, and noting that $\left\| Em_0^{(n)}(T) \right\| = 1$, leads to:

$$\left\| Em_\delta^{(n)}(T) \right\| \le n^*\delta + 1,$$

where n^* is related to n according to the noise-budget analysis appropriate to the form of Emanator implementation. When n refers to the 137 independent terms, each of max norm 1, comprising each Trigintaduonion emanation, n^* is the familiar $1/\alpha$. When n refers to the 29 independent dimensions of emanation along a particular chiral emanation (with perturbations), then n^* is the effective number of dimensions in the iterative mapping that results. This is needed in the Results that follow to get the $\{\alpha, \pi \, c_\infty\}$ relation.

From the form above, application of the Kato-Rellich theorem is equivalent to a noise budget analysis to arrive at the same inequality. We inject an amount of noise δ in the form of the emanation chosen and determine the amount of noise, worst-case, that might present after the chosen emanation operation is performed. Since the terms are all max norm 1, this decomposes into a simple counting on the number of independent terms, free dimensions, etc. This allows for a straightforward counting process to arrive at a number of solutions as will be shown in the results that follow.

3.2.4 Trigintaduonion Emanation: achirality from sum on all chirality
If analyticity confers Laplace's equation, from electrostatics for example, what may confer electrodynamics? For this we need something that can be dynamical and in the current theory of trigintaduonion emanation (projection) we only discussed how to connect to a 10D emanation with alpha perturbation into 32D, now complexified to 64D (surmised to have been there in the projection to the perturbed 32D state at the outset, with the initial trigintaduonion projection emergence). In practice, there are four chiralities, and for a given chirality (with unit norm) there are 29 dimensions of freedom (10D + 19D of chirally consistent perturbation). When analytic extension is taken to give maximal information flow, the effective dimension for each of the four chiralities is 29* (detailed in [37]). This clear decomposition into 29* independent effective dimensions is then revealed in the $\{\alpha, \pi, C_\infty\}$ relation in [37]. The Mandelbrot Set is one of many that encounter the universal constant C_∞. The Mandelbrot set also describes a 2D fractal boundary at its "edge of chaos". If driven to similar optimality in approaching a zero-value (a zero-divider issue), we see a two-value zero-crossing specification effectively like a double zero. The parameterization of the zeros of the Emanator at chiral zero-divisor points will thus be as double-zeros.

Recall the description of the emanator from [37]:

$$T_{chiral}^{(k)} = \begin{cases} ((0,\alpha),\beta) \\ ((\alpha,0),\beta) \\ (\beta,(0,\alpha)) \\ (\beta,(\alpha,0)) \end{cases}, where \; T_{chiral}^{(k)} = 1 + i\delta.$$

$$\text{Emanation}(\mathbf{T}) = \frac{1}{N} \sum_{k \in \{4x72\}^n} \mathbf{T} \bullet T^{(k)}_{chiral} = \frac{1}{N} \sum_{K \in 4\ chiralities} \mathbf{T} \bullet \overline{T}^{(K)}_{chiral}$$

Suppose we add the rule that emanation may not proceed when a particular chirality is zeroed-out, in other words:

$$\mathbf{T} \bullet \overline{T}^{(K)}_{chiral} \neq \mathbf{0}.$$

For 'normal' numbers this goes without saying, since for real numbers if we have $r_1 \times r_2 = r_3$ then $r_3 \neq 0$ $if\ neither\ r_1 = 0\ or\ r_2 = 0$. This holds true for the Real, Complex, Quaternion, and Octonion numbers. This does not hold true for Sedenions or higher. For sedenions the dimensionality of the zero-divisor event is mostly constrained, while for trigintaduonions it is significant. If such zeros were eliminated from the emanator description by using analytic extension component-wise (on 29* effective components) we see how a description devoid of matter (pure static field with no source or sink) might acquire matter by way of extending to a maximal domain of analyticity be removing zero-divisor events (a Wick transformation from real dimensionless action to pure imaginary action that is dimensionless but consisting of a dimensionful ratio). For what follows, let's parameterize the zero-divisors and index them such that:

$$\mathbf{T} \bullet \overline{T}_{chiral}(S^*_i) \to \mathbf{0}\ as\ double\ zero\ \forall S^*_i .$$

Chapter 4. Achiral T-emanation has 29* effective dimensions

4.1 Introduction
In this chapter we obtain an estimate of the effective dimension of information transmission in an achiral T-emanation process.

4.2 The construction of an achiral emanator
Consider the emanator described in the previous section: $(\tilde{A},\beta) = ((\tilde{a},\alpha), \beta)$. Let's shift to representing the full octonion part by O: $((O,\alpha), \beta)$. There are four types of chiral emanation:

$$T^{(k)}_{chiral} = \begin{cases} ((O,\alpha),\beta) \\ ((\alpha,O),\beta) \\ (\beta,(O,\alpha)) \\ (\beta,(\alpha,O)) \end{cases}, where \ T^{(k)}_{chiral} = 1 + i\delta.$$

From unit norm we have $\alpha^2 = 1 - O^2 - \beta^2$, with $\pm$ sign choice on α, similarly for β. The constraint to near-unity emanation will be dropped in later analysis (the split Cayley formulation).

Suppose we have a unit norm base trigintaduonion as before, but let's now attempt to construct an achiral form of emanation from the set of four types of chiral emanations (and summing over the four sign conventions for $\pm\alpha$ and $\pm\beta$).

$$\text{Emanation}(\mathbf{T}) = \frac{1}{N} \sum_{\{k\}} \mathbf{T} \bullet T^{(k)}_{chiral},$$

where $(1/N)$ is a normalization (to recover unit norm) and the Results show analysis of this emanator as well as the chiral emanators $T^{(k)}_{chiral}$ individually.

A straightforward perturbative analysis, or noise budget analysis, can be used on the above emanator to determine the effective number of dimensions in the iterative mapping corresponding to the emanation step. This is a construction for dimensional regularization (another analyticity argument, used in QFT renormalization [45]), and it shows that the achiral emanation definition above is on the right track, but has not differentiated within the four chiralities properly. Consider the first chiral emanation family with the template $((O, \alpha), \beta)$. In the achiral emanation by simply summing over each of the four chiralities, the emanator for a given chirality is generated randomly and according to the indicated template, where all seven imaginary components of the octonion O have a small perturbative contribution. Let's now consider 14 possible perturbations within a given "suit" (chirality) from the pure imaginary octonion modulations (positive or negative). For the $((O, \alpha), \beta)$ chirality this corresponds to generating emanators with the form: $O = (\sim 1, 0, 0, \dots \delta, \dots 0)$, with δ perturbation in each of the seven positions and then further divided according to whether it is positive or negative. Let's also consider 4 additional perturbations according to the template when $\hat{\alpha}^2 = 1 - O^2 - \beta^2$ and the template has perturbation (positive or negative) at the $\hat{\alpha}$ component: $((O, \hat{\alpha} + \delta), \beta)$. Similarly for $((O, \alpha), \hat{\beta} + \delta)$. Thus, each chirality is split into 18 subtypes, and for the four chiralities this results in 4x18=72 terms in the emanator sum, with 4 separate sums on the 72 according to the $\pm \alpha$ and $\pm \beta$ conventions on the template. Thus

$$\text{Emanation}(\mathbf{T}) = \frac{1}{N} \sum_{k \in \{4x72\}} \mathbf{T} \bullet T^{(k)}_{chiral},$$

This is referred to as the Emanator with the 72-card deck and in the results it will be shown to provide a relation between the fine structure constant, π, and the Feigenbaum Universal bifurcation parameter C_∞, that is correct to the highest level of experimental and theoretical precision known on the fine structure constant.

If we only consider the 14 subtypes from pure imaginary octonion contributions, there are 4x14=56 card types. Respective to a particular chiral template, there are 22 zero-positions from the imaginary octonion sector with α (7 components) and the imaginary sedenion sector associated with β (15 components), giving rise to 22 chiral propagators of the form $\mathbf{T} = (\sim 1, 0, 0, \dots \delta, \dots 0)$. If we combine the 56 minor subtypes or 'cards' and the 22 major cards, we arrive a similar complete system of perturbations, whose sum would again be achiral. This latter case, with a 78 card deck, is referred to as "Tarot Emanation" due to the similarity to the Tarot deck with 56 minor arcana and 22 major arcana. For the derivation to follow in the Results, however, the 72 deck is most accessible to analysis. Further understanding of the multiple-multiplication 'paths' in an achiral path sum of chiral emanations is left to the discussion.

4.3 The relation of C to α (and thus π)

Recall that we have achiral emanation in the form:

$$\text{Emanation}(\mathbf{T}) = \frac{1}{N} \sum_{k \in \{4x72\}} \mathbf{T} \bullet T^{(k)}_{chiral},$$

Each of the chiral trigintaduonions has a template of fixed parameters, involving 3 of its 32 dimensions, leaving 29 dimensions 'free'. The effective dimension will be 29 plus a correction due to imaginary contributions to the noise transmission with each chiral multiplication. Consider a noise, or perturbation, contribution δ, in generating the chiral emanators of the various types. From a base trigintaduonion with chiral multiplication in the Emanator sum, for the 29 'free' dimensions respective to that chiral multiplication path, we have noise transmission for each of the independent 72 elements from the 72-deck sum, assume worst-case noise transmission into each of the independent emanator sum terms (72) in each of the 29 dimensions, whose imaginary component is again maximal antiphase at maximum noise transmission, thus $(\pi/29x72)$ for each of these terms. Thus, as a conditioning step, consider a trigintaduonion resulting from Emanation with a 72-deck as described, with noise in each free dimension going as $\delta(1+i(\pi/72)/29)$. Now let's consider this noise transmitted through a general emanation step:

The real part, δ, will transmit to δ in the new trigintaduonion, but since the emanation process uses a 'deck' of 72 valid chiral emanation types, a correction is needed since 3 of these emanation types are not valid for the real emanation path (the 3 chiral emanation that have α or β at the T[0] position are locked into α positive (~1) or β positive (~1), respectively, thus exclude 3 of the 4 {±α,±β} cases). This amounts to the real part $\delta \rightarrow$ $\delta(1 - \frac{1}{29}\left(\frac{3}{72}\right) 4(\frac{\pi}{72})(\frac{\pi}{137x29})\hat{\imath}$), where the correction on the real part is (3/72) of the $\delta(\pi/72)/29$ imaginary part transmitting as a new noise factor δ', where there are four transmission chiralities (each with its own resulting imaginary, so sum to 4 after renormalization, unlike the real part where "1" is the same in the four chiral sums, thus normalization, divide by four at this stage, reduces to "1" for the real part shown). For convenience, the four different imaginary values are summed as the 4i shown. When considering effective dimensions later this will be valid when linear additivity is assumed (not adding in quadrature). The modified noise factor $\delta'' = \frac{1}{29}\left(\frac{3}{72}\right)4\left(\frac{\pi}{72}\right)$ described thus far, is then multiplied, in the emanation product, by the maximal noise imaginary component allowed in the 137 independent terms in the 29 independent dimensions, $\theta = (\pi/29\text{x}137)$, thus the form shown.

The imaginary part, $\delta(i(\pi/72)/29) \rightarrow 4\delta((\pi/72)/29)\hat{\jmath} + 4\delta((\pi/72)/29)\left(\frac{\pi}{72}\right)\left(\frac{\pi}{137x29}\right)\hat{k}$, where the first term simply results from the $\delta(i(\pi/72)/29)$ noise injection hitting the ($\sim$1) real component in the chiral triginataduonion multiplication, again with a factor of 4 from the 4 separate chiral sums. The second term has the $\delta((\pi/72)/29)i$ noise injection factor, the 4-factor, as before, and a $(\pi/72)j$ factor for the 72-deck chiral emanations and within that a $(\pi/29\text{x}137)k$ factor respective to a particular chiral emanation. Again, the ijk imaginary products for different i, j, k's, is all grouped as $\hat{k}$.

Now to multiply the noise for one of 29 free dimensions by 29 and sum the magnitudes of the real and imaginary components. Dividing out the noise injected δ, we thereby arrive at an expression for the effective dimensionality as seen by noise transmission:

$$\text{Dim effective} = 29 + (4\pi/72)[(1+\theta\{(\pi/72)+(3/72)\}]$$

Thus, we expect the maximum perturbation amount α, when inverted, to be related to the Feigenbaum bifurcation constant according to the number of effective dimensions:

$$\alpha^{-1} = (c_\infty)^\gamma = 137.035999206...,$$

where $\gamma = (1/2)(29 + (4\pi/72)[(1+\theta\{(\pi/72)+(3/72)\}])$, and $\theta = \pi/(137\text{x}29)$.

4.4 Re-analysis of achiral emanation with explicit use of Exp Map

There are 4 chiralities, so to get an achiral emanator candidate, minimally need a "4-card deck" to emanate in the four chiralities, with emanator equal to normalized sum. The actual deck appears to require a normalized sum over sub-chiralities, as will be explicitly enumerated in what follows.

Here are the four chiralities with real fluctuation noise shown:

$$((\,(\,(O[0] \pm \delta, \ldots\,), \alpha \pm \delta), \beta \pm \delta)$$
$$((\,\alpha \pm \delta, (O[0] \pm \delta, \ldots\,)), \beta \pm \delta)$$
$$(\beta \pm \delta, (\,(O[0] \pm \delta, \ldots\,), \alpha \pm \delta)\,)$$
$$(\beta \pm \delta, (\,\alpha \pm \delta, (O[0] \pm \delta, \ldots\,)\,)\,)$$

where α is a real octonion and β is a real sedenion, and Tem is an equal weight sum of the action of each of the sub-chiral propagations on the base T, with the fluctuations indicated each done separately. We have the constraints $\alpha \neq 0$, $\beta \neq 0$, and common octonion O not pure real.

Each of the δ's is an independent fluctuation corresponding to its own sub-chiral emanation, but no subscripting on δ's is used or shown. There are thus 9x2x4=72 independent *imaginary* noise fluctuations to consider in the exp(Im(Tem)) evaluation (that automatically provides unit-norm). The real noise fluctuations in the real (first) component are, thus, not counted. If our definition for Tem entails only one card being dealt, then the sum over those possibilities is the sum

$$\mathbf{T} \bullet \mathrm{T}_{\mathrm{em}} \equiv \mathrm{Emanation}(\mathbf{T}) = \frac{1}{72} \sum_{k \in \{72\}} \mathbf{T} \bullet T^{(k)}_{chiral}$$

For one-card, or a one-step, emanation, with real components and real noise, this makes sense from the counting shown, and it's what we use going forward. Using this will allow an entirely separate method for evaluating α (here at the one-card hand approximation). This will be done by determining the effective dimension 29*>29 of maximal information propagation (or maximal noise fluctuation). Before moving on, however, let's examine what happens when we allow complex noise fluctuations as this will trivially be allowed when we consider $C \times T$ via exp(iT) in later discussion anyway.

Maximum information transmission involves a complex extension to the T components and their noise fluctuations, but in doing this it must retain emanation structures such as the octonionic triple that occurs in previous expressions (starting with the proof of the T_{chiral} solution itself), which leads to the counting that gives 137 independent terms, etc. Thus, the maximal complex extension on the noise is that it remain real in the octonion components:

$$(((O[0] \pm \delta, ...), \alpha \pm i\delta), \beta \pm i\delta)$$
$$((\alpha \pm i\delta, (O[0] \pm \delta, ...)), \beta \pm i\delta)$$
$$(\beta \pm i\delta, ((O[0] \pm \delta, ...), \alpha \pm i\delta))$$
$$(\beta \pm i\delta, (\alpha \pm i\delta, (O[0] \pm \delta, ...)))$$

The first chiral T component is where new imaginary terms might arise (the others are already counted since in imaginary components). We see there are six more, so the deck is now78.

When to use the 72-deck and when to use the 78-deck isn't clear yet, this will eventually be something that can be determined by how the theory converges on the non-approximate α^{-1} given above. At one card, or the first card, emanation is just a sum over chiral T's, so still a T product (acting on T_{base}) without the complication of zero divisors (to be described in later sections), so this is a convenient dividing line between the α estimation, or theory value, based on T, and that based on $C \times T$. The $C \times T$ description, trivially allowing complex noise fluctuations, may go best in the multi-card emanation description where the exponential map $\exp(iT) = C \times T$ will be indicated anyway. Further consideration of emanation with a multi-card hand (or multi-step path) will be discussed at a later time.

All noise terms will be treated additively, including terms in different imaginary components as well as imaginary noise terms in the real component. The criterion for max noise (in-phase constructive interference) gives the extreme of linear additivity. (Not like Gaussian statistical noise that adds in quadrature.) Also note that the discussion in terms of "noise transmission" and "information transmission" will be used almost interchangeably, whenever one description or the other best suits the analysis it will be used. Note that with this kind of noise analysis we can effectively shift around T noise terms associatively. Also note that application of the Kato-Rellich theorem [15,37] is related to the noise budget analysis done here focusing on first order terms.

There are 137 independent tri-octonionic terms in each of 29 free components indicated by a particular chirality (within the 32 components of a general trigintaduonion). This is a nontrivial result since ($T_{chiral} \bullet T_{chiral}$) is no longer T_{chiral} type (but still T_{norm1} type), so direct expansions are needed to identify the number of independent terms and this is briefly described below, with more detail in [6].

4.5 Single-step achiral 72-deck emanation has noise propagation dimension 29* [46]

Obtaining an achiral emanation from a collection of chiral emanations requires that all chiralities be summed over (there are four) as well as sub-chiralities (there are 72). Noise analysis requires collecting of first-order terms. Analysis of noise transmission indicates 29* dimensions, where:

$$29^* = 29 + \left(\frac{4\pi}{72}\right)\left[1 + \left(\frac{\pi}{137 \cdot 29}\right)\left(\left(\frac{\pi}{72}\right) + \left(\frac{3}{72}\right)\right)\right]$$

The above result was obtained in [37] to describe the 72-card chiral 'deck' of chiral emanation products for a single-step emanation. In the Methods to follow this is reviewed and elaborated further.

4.6 The effective noise transmission dimension

For emanation we produce an achiral sum from the four primary chiral emanations. To achieve an emanator with counting that provides agreement with the $\{C_\infty, \alpha, \pi\}$ relation, we see the need for sub-achirality as well, necessitating a sum over the 'full deck' of specific subtypes of chiral emanation. One of the chiralities is written $T = ((O, \alpha), \beta)$ where the normalization factor to achieve unit norm is suppressed, where O is an octonion, α is a real octonion, and β is a real sedenion, or a real octonion in the pair S=(β,0), depending on notational convenience. Also, depending on notational convenience, often α and β will be treated as simple real numbers (equal to the real component of the respective octonion).

Let's consider the generative process of arriving at an achiral emanation, and the counting results that are obtained:

(1) There are four chiral forms, let's start an emanation by choosing a common octonion element for all four chiralities. Now let's choose $[+\alpha, +\beta]$. We can't have $\alpha = 0$ or $\beta = 0$, so we have a splitting into four disconnected regions in the $[\alpha, \beta]$ plane (the four quadrants). Let's choose positive $[\alpha, \beta]$ then generate four cases (respective to the four quadrants) by having cases with $[\pm\alpha, \pm\beta]$. (In this way the four principle chiralities will have the same normalization constant).

(2) Each chiral form has 9 imaginary components, Let's consider two maximal perturbations for each (via α perturbation limit), one with the

imaginary component plus the max-α fluctuation, one with minus the max-α fluctuation. The normalization constants will now differ.

The number of Cases = (4 chiralities) x (4 [$\pm\alpha, \pm\beta$] subchiralities) x (9 imag perturbations) x (2 max pert. at $\pm$max-α) = 4 x 72. Consider the 4 x 72 cases summed over all but the 4 chiralities. We may have 'merging' at the core four chirality level:

$$\text{Emanation}(\mathbf{T}) = \frac{1}{4} \sum_{k \in \{4\}} \mathbf{T}_{base} \bullet T^{(k)}_{chiral},$$

Going further, and using the notation of a simple product relation will be adopted for the "1 card" emanation. It may not exist, from the norm sum form, but in the noise analysis, with the simple rules on counting noise terms additively, it helps to arrange terms for that counting, Thus, let's consider the noise in:

$$\mathbf{T}_{base} \bullet \mathbf{T}_{em,(1\ card)}$$

which will have groupings:

$$\{real + real\ noise + imag\ noise\}$$
$$\times \{(real \approx 1) + real\ noise + imag\ noise\}.$$

From the previous analysis of the maximal noise with the achiral emanator we had the "noise triangle" with noise magnitude Hc, noise angle $\theta = \pi/(137 \times 29)$, imaginary noise = $(Hc)\sin(\theta)$, real noise = $(Hc)\cos(\theta)$, where $H = 137/\cos(\pi/137)$ and $c = \theta/\sin\theta$. Let's take $\mathbf{T}_{base}$ to have this generic form, associated with maximal information flow α^{-1} = max real noise = $(Hc)\cos(\theta)$. Also, we don't know the overall scale, so common scale factors can be dropped (H), such that the imaginary noise term is $\delta\theta i$, and there are 29 of them for the 29 unconstrained dimensions (due to choice of chirality and norm=1):

$$\{1, ...29\delta\theta i\} \times \{real + imag + real\ noise + imag\ noise\}.$$

For the emanator T on the right-hand-side, we know the imaginary terms will be the same, thus a $29\delta i$ factor. Note, the analysis is done for the first quadrant of the [$\pm\alpha, \pm\beta$] subspace, the one connected to T=1 with [$+\alpha, +\beta$] such that its real term is $T_{em}[0] = 1 - \Delta$. (Later, the four [$\pm\alpha, \pm\beta$] sub-chirality sectors will give an overall factor of four.) Aside

54

from the $29\delta i$ factor, the imaginary terms will have a max 1 transmission path as well as an equipartition of max-antiphase π over 29 free dimensions and 137 independent terms in each, thus have a factor of $[1 + \theta i]$ (ignoring a common normalization term that factors out in what follows). So far we've described the imaginary noise that might occur 'internal' to a chiral sum, effectively for one card of the deck (or for a minimal 4-suit achiral that expresses the key 29 free dimensions property). For the imaginary noise terms we expect another phase factor from the 'deck sum' (here the phase analysis on the 137 independent terms is separated from the phase analysis, at deck-level, of the 72 independent cards). This factor should involve an equipartition over 72 cards, each with 29 free dimensions: $\pi/(72 \times 29)$. Putting this together:

$$\{1, \ldots .29\delta\theta i\} \times \left\{1 - \Delta, 4 \times 29\delta i \times \frac{\pi}{72 \times 29} \times [1 + \theta j]\right\}.$$

Simplified further:

$$\{1, \ldots .29\delta\theta i\} \times \left\{1 - \Delta, \delta\left(\frac{4\pi}{72}\right)i + \delta\left(\frac{4\pi}{72}\right)\theta j]\right\},$$

where the i and j different imaginary terms are only held separate for clarity, eventually they will be merged as one imaginary term (according to the additivity on terms in the maximum noise analysis). Of the cards in the 72-deck, 3 of them don't match with the choice of T$\cong$1 with $[+\alpha, +\beta]$ subspace. This is why in the noise analysis we have $T_{em}[0] = 1 - \Delta$ and not simply '1' (when ignoring overall normalization factor), the Δ corrects for needing to subtract 3 cards from the emanation deck, and thus their channels of (card-level) noise. Thus, we have:

$$\Delta = \left(\frac{3}{72}\right)\delta\left(\frac{4\pi}{72}\right)\theta j.$$

The maximum noise transmission then occurs with:

$$\{1, \quad 29\delta\theta i\} \times \left\{1 - \left(\frac{3}{72}\right)\delta\left(\frac{4\pi}{72}\right)\theta j, \quad \delta\left(\frac{4\pi}{72}\right)i + \delta\left(\frac{4\pi}{72}\right)\theta j]\right\},$$

The imaginary noise terms, at first-order, are:

$$\delta i(29)\left[29 + \left(\frac{4\pi}{72}\right)\{1 + \left(\frac{\pi}{137x29}\right)[\left(\frac{\pi}{72}\right) - \left(\frac{3}{72}\right)j]\right]$$

Gathering the imaginary noise terms, *but all additive*, such that maximum noise is achieved with whatever constructive interference on noise phase, we get:

- maximum imaginary noise
  $= \delta\{ 29 + (4\pi/72)[(1+\theta\{(\pi/72)+(3/72)\}] \}, \theta = \pi/(137 \times 29).$
- effective noise transmission dimension
  $= \{ 29 + (4\pi/72)[(1+\theta\{(\pi/72)+(3/72)\}] \} \equiv 29^*.$

4.7 'Edge of chaos' maximal perturbation hypothesis [37]

Consider the 'edge of chaos' maximal perturbation in each of the 29^* dimensions to be at position C_∞ (see Appendix for background on Mandelbrot Set), which is on the negative real axis, i.e., at π rotation to have -1 factor, ***thus at maximal antiphase***. This results in the relation for maximal perturbation at maximal antiphase (maximum reference angle with sign chosen positive by convention) has a lower bound on α given by:

$$\alpha_0^{-1} = \left(\sqrt{C_\infty}\right)^{29^*}.$$

where

$$C_\infty = 1.4011551890920506004 \ldots$$

This ties $1/\alpha$ to the second Feigenbaum constant C_∞ in the context of the Mandelbrot set. It is well known that the Feigenbaum constants are universal, and part of a description of a universal transition to chaos regime. The Mandelbrot set is also universal [38], and maximal in that its fractal boundary has maximal fractal dimension of 2 [38], a detail that will be important in the meromorphic matter description given later.

For C_∞, most references only provide $C_\infty = 1.401155189 \ldots$, and a higher precision tabulation is not readily found, so use is made of the relation

$$C_n = a_n(a_n - 2)/4,$$

together with the tabulation on a_∞ [47]:

$$a_\infty = 3.56994567187094449018 \ldots$$

The resulting C_∞ is:

$$C_\infty = 1.4011551890920506004 \ldots$$

The resulting α_0^{-1} is:

$$\alpha_0^{-1} = 137.03599933370198263 \ldots$$

For the multi-card analysis we have:

$$\alpha^{-1} = \alpha_0^{-1} + \alpha_1^{-1} + \cdots$$

where α_0^{-1} involves the sum over emanation by one-card. For sum on two-chiral products we have further 'noise' contribution α_1^{-1}. With the multi-card modifications, albeit small, there is the complication of shift

56

from 72-deck to 78-deck, and whether there is a chiral step ('card') type that can be exactly repeated (i.e., are cards from the deck played with replacement when considering a multi-card flop sequence). There may be a reason why the sums must be done without card-replacement. This might be because card replacement would allow degenerate tri-octonionic product terms, again throwing off the 137 braid term total, perhaps, leading to non-optimality. This is being explored in further work and will not be discussed further at this time.

4.8 Trigintaduonion Emanation: achirality from chirality

The Mandelbrot Set (see Appendix) is one of many that encounter the universal constant C_∞. The Mandelbrot set also describes a boundary with 2D fractal dimension [38] at its "edge of chaos" [37]. If driven to similar optimality in approaching a zero-value (a zero-divider issue), we see a two-value zero-crossing specification effectively like a double zero. The parameterization of the zeros of the Emanator at chiral zero-divisor points will thus be as double-zeros.

For what follows we use the simple description of the emanator:

$$T^{(k)}_{chiral} = \begin{cases} ((0,\alpha),\beta) \\ ((\alpha,0),\beta) \\ (\beta,(0,\alpha)) \\ (\beta,(\alpha,0)) \end{cases}.$$

$$\text{Emanation}(\mathbf{T}) = \frac{1}{N} \sum_{k\in\{4x72\}^n} \mathbf{T} \bullet T^{(k)}_{chiral} = \frac{1}{N} \sum_{K\in 4\ chiralities} \mathbf{T} \bullet \overline{T}^{(K)}_{chiral}$$

If working with non-split T's, then we restrict to emanations that are perturbations of unity:

$$T^{(k)}_{chiral} = \begin{cases} ((0,\alpha),\beta) \\ ((\alpha,0),\beta) \\ (\beta,(0,\alpha)) \\ (\beta,(\alpha,0)) \end{cases}, where\ T^{(k)}_{chiral} = 1 + i\delta.$$

From unit norm we have $\alpha^2 = 1 - 0^2 - \beta^2$, with $\pm$ sign choice on α, similarly for β.

If working with split T's (bi-sedenions, etc.), then we have manifest Lorentz Invariance (shown in Ch. 7). So, it is often convenient to work with split-T' since this is manifest from the outset.

57

Chapter 5. Achiral T-emanation leads to the Standard Model, "small h", and meromorphic matter

We now show that achiral T-emanations can be put into the form that indicates a gauge theory of precisely the form found for the Standard Model of Particle Physics.

5.1 Exp{i Em(T)} has the gauge of the Standard Model
5.1.1 *Emanation*$(T) \cong O \times H$

The simplest emanation, at the (summed over) chiral level, is a mixture of four forms:

$$T^{(k)}_{chiral} = \begin{cases} ((0,\alpha),\beta) \\ ((\alpha,0),\beta) \\ (\beta,(0,\alpha)) \\ (\beta,(\alpha,0)) \end{cases}.$$

Consider the chiral emanation $((0,\alpha),\beta)$ if repeated in right operation on **T**, with $\alpha = 0$ and $\beta = 0$ (not an allowed degenerate case in emanation sum):

$$\left(\left(\mathbf{T} \bullet ((\mathbf{0},\alpha),\beta) \right) \bullet ((\mathbf{0},\alpha),\beta) \bullet \bullet \bullet \right)$$

becomes

$$\left(\left(\mathbf{T} \bullet ((\mathbf{0},0),(0,0)) \right) \bullet ((\mathbf{0},0),(0,0)) \bullet \bullet \bullet \right)$$

where the octonion product with **O** is carried directly to component level in **T**. Let T=((x,y),(u,v)), then get T=((((x**O**)**O**...**O**), ((y**O***)**O***...**O***)), (((u**O***)**O***...**O***), ((v**O**)**O**...**O**))). Thus, have an octonion right product, repeated in each of the four octonions in the T, giving the action of the T in this case to be like that of a common octonion right product. A sequence of octonion products produces a SU(3) algebra, and SU(3)xU(1) with suitable normalization [20,23]. If we now consider the actual chiral form $((0,\alpha),\beta)$ without $\alpha = 0$ and $\beta = 0$, is it possible to show an additional SU(2)$_L$ (and SU(2)$_R$) algebraic product to arrive at the Standard Model (as well as a possible SU(2)$_R$ extension)? The answer is Yes. And this is easy to show because the main result derives from the property that elements of the algebra C × H x O, when taken as a repeated product, have a associative product group symmetry that is that of the Standard Model: U(1) × SU(2)$_L$ × SU(3) [23].

In the noise perturbation analysis we already generalized to complex noise on the basis that later applications would involve sums over trigintaduonion phase terms consisting of exponentiated iT_{em}. Fortunately the exp map is well-defined for hypercomplex number generalizations [23] and, via the generalized De Moivre relation, expresses a clear 'lift' from a T number to a C × T number. What remains is to show that the particular form of T_{em} chosen is manifestly **O x H**. As suggested earlier, this is the case when the emanator sum has sub-chiralities generated from a common octonion. This leads to expressing $T_{em} =$ $(O_{em}a, O_{em}b, O_{em}c, O_{em}d) = O_{em} \times H$, where $H = (a, b, c, d)$.

5.1.2 Em(T), the standard model of particle physics, and quantum matter

The chiral trigintaduonions T, with right product operation ((T x T) x T)..., used previously for maximum information transmission, are here shown to be H×O when arranged for achiral emanation. When considering a sum over chiral emanations to obtain an achiral emanator, with T as phase factor, we have the exponentiation operation exp(iT), which leads to a theory that is C × H × O. As such, we have the foundation for the associative operator algebra of the Standard Model: U(1) × SU(2)$_L$ × SU(3). A complication with T products is you can have zero divisors. A framework is adopted to remove the zero divisors by requirement of maximum domain of analyticity on the log trigintaduonion multiplication, resulting in a description for the meromorphic precipitation of matter. In this process a fundamental quantum is indicated from the zero-divisor residue terms. Analyticity in the form of a Wick rotation also provides a mechanism whereby we can transition to a dimensionful action and quantum and arrive at an explanation for the critical 'smallness' of Planck's constant. The emanator is $T_{em} \cong$ H×O, and provides a description of a possible meromorphic origin for point-like matter.

Getting an associative algebra from the repeated operation of a non-associative algebra is first described in [20] in the context of repeated octonion products: ((OxO)xO)..., where the algebra SU(3) can result. This is found to be equivalent to fixing one of the octonion imaginary components in such a repeated-product operation [20,23]. Dixon shows in

[23] that the $C \times H \times O$ product algebra lays the foundation for the associative operator algebra of the Standard Model: $U(1) \times SU(2)_L \times SU(3)$. In later work this is explored in the form of ideals [24]. Emanator theory, described in a brief background to follow, is based on unit-norm propagation at maximal dimension. It turns out this maximal dimension is not octonion-based but trigintaduonion-based [1], although it does have an octonion sub-algebra: $T_{em} \cong H \times O$, as will be shown here.

In prior work [1,6,15,16,22,37,46] we hypothesized maximal algebraic information flow, where the "emanation" of information is represented as multiplication by an element of an algebra in two steps: (i) take the maximal current-state element that is a unit-norm trigintaduonion; and (ii) perform the emanator step that consists of an achiral sum of multiplications with chiral trigintaduonion emanators. In prior efforts this was considered without the complication of zero divisors. We will see that zero divisors act as "sources", so the prior work was effectively analysis of sourceless information flow.

In [22] we show $T_{em} \cong H \times O$ (which will lead to the standard model) and we consider zero divisors and their impact on the maximal information flow and in doing so see a mechanism for meromorphic precipitation of quantum matter with dimensionful action. When we consider $\exp(iT_{em})$, we get a mathematical object that is $C \times H \times O$, thus whose product algebra becomes $U(1)xSU(2)xSU(3)$ according to Dixon [23]. If this route to the standard model is unsatisfactory, consider that the T emanation sum pre-normalization can be grouped as a 2x2 real matrix (for split form) that is finite, thus SU(2) with normalization: $T_{em} \cong [2x2$ real$] \times O$;
where repeated steps of
$$\exp(iT_{em}) \rightarrow \text{unit norm } Cx[2x2 \text{ real}]xO$$
re-group as repeated steps of CxCxCx..., and [2x2 real]x[2x2 real]x..., and OxOxO...., with unit normalization, thus the product algebra is
$$U(1)xSU(2)xSU(3).$$

5.2 Maximal Information Emanation (MIE)
5.2.1 Achiral T-emanation with maximal domain of analyticity via removal of zd singularities – Meromorphic Matter
The success of the 29* factoring argument (getting an estimate of α good to 10 decimal places) suggests that such a factoring of the emanation process is valid in certain applications. Let's consider such a factoring together with addition of complex structure and examine the local

61

analytic domains that result (29* of them). Why would we do this? Well suppose your algebraic formalism has a critical weakness, such as the existence of zero divisors. How might a maximum information theory go about repairing the situation? In a particular complex dimension we could define a fundamental "log T" theory, where the singular zero must be removed, with resulting residue… and this is what is done, with some interesting results in what follows.

5.2.2 The zero-divisor problem

For the real numbers, $xy=0$ only if $x=0$ and/or $y=0$, i.e., there are no "zero divisors: $y=0/x$. This is true of all division algebras (the Cayley algebras R, C, H, O). Starting with the sedenions (S), and even more so with the trigintaduonions (T), we have zero divisors (see Appendix). Let's re-examine the emanator product with this complication in mind.

Suppose we add the rule that emanation may not proceed when a particular chirality is zeroed-out in other words:

$$\mathbf{T} \bullet \overline{T}^{(K)}_{chiral} \neq \mathbf{0} \, .$$

For sedenions the dimensionality of the zero-divisor event is mostly constrained, while for trigintaduonions it is significant (see Appendix in [22]). If such zeros were eliminated from the emanator description by using analytic extension component-wise (on 29* effective components) we see how a description devoid of matter (pure static field with no source or sink) might acquire matter by way of extending to a maximal domain of analyticity be removing zero-divisor events (a Wick transformation from real dimensionless action to pure imaginary action that is dimensionless but consisting of a dimensionful ratio). Using the notation adopted in [46], let's parameterize the zero-divisors and index them such that:

$$\mathbf{T} \bullet \overline{T}_{chiral}(S^*_i) \rightarrow \mathbf{0} \ as \ double \ zero \ \forall S^*_i \, .$$

5.2.3 Emanation when base trigintaduonion contains Zero Divisors [46]

Consider emanation when the base trigintaduonion is pure imaginary, and thus a potential zero-divisor $\mathbf{T_{ZD}}$:

$$\text{Emanation}(\mathbf{T_{ZD}}) = \frac{1}{N} \sum_{k \in \{4x72\}^n} \mathbf{T_{ZD}} \bullet T^{(k)}_{chiral} \, ,$$

and suppose the number n (like the number of cards in 'flop' to make a reading) is such that $\{4x72\}^n$ is large, such that the sum on trigintaduonion products is dominated by stationary phase terms. Such domination by stationary phase is expected with appropriate handling on the normalization, even without zero real component and unit norm, since we have phase addition on a compact space, the 31-sphere. We now have a new mechanism driving the stationary phase solution, however, due to the existence of zero divisors, for which a new type of solution class is indicated. Suppose stationary phase in this context selects such that:

$$\text{Emanation}(\mathbf{T_{ZD}}) = \frac{1}{N}\mathbf{T_{ZD}} \bullet (R + \mathbf{T_{ZD*}}) = \mathbf{T_{ZD}}, \qquad \Delta\mathbf{T}_{base} = 0$$

where $\mathbf{T_{ZD*}}$ is the zd conjugate to $\mathbf{T_{ZD}}$, i.e. $\mathbf{T_{ZD}} \bullet \mathbf{T_{ZD*}} = 0$, and N is the appropriate normalization constant to arrive at unit norm as before. Since $\Delta\mathbf{T}_{base} = 0$, in the emanation process it is unchanging, thus this is the condition that will relate to the classic equilibrium (or quantum stationarity).

Let's now consider the $\mathbf{T}_{base}$ that consists of a sum over a countable collection of zero divisors with separate weighting factors:

$$\mathbf{T}_{base} = \sum_{i \, \epsilon \, all} a_i \mathbf{T}_{ZD,i}$$

Suppose stationary phase in this context selects such that:

$$\text{Emanation}(\mathbf{T}_{base}) = \frac{1}{N}\mathbf{T}_{base} \bullet \sum_{i \, \epsilon \, all} \mathbf{T}_{ZD*,i} \bullet (\mathbf{T}_{ZD*,i})^{-1} = \mathbf{T}_{base}$$

where the order of 3-T multiplications is with the inverse last, and where an overall constant is eliminated by the renormalization term to arrive back at the starting base trigintaduonion. This appears to be the general condition for describing the emanation form of equilibrium.

Let's now consider what happens if the real component is nonzero as well (and assume Z ZD's):

$$\mathbf{T}_{base} = R + \sum_{i \, \epsilon \, all} a_i \mathbf{T}_{ZD,i}$$

$$\text{Emanation}(\mathbf{T}_{base}) = \frac{1}{N}(R + \sum_{i \, \epsilon \, all} a_i \mathbf{T}_{ZD,i}) \bullet \sum_{i \, \epsilon \, all} \mathbf{T}_{ZD*,i} \bullet (\mathbf{T}_{ZD*,i})^{-1}$$

$$\text{Em}(\mathbf{T}_{base}) = \frac{1}{N}\left(Z\mathbf{R} + \sum_{i \in all}(Z-1)a_i\mathbf{T}_{ZD,i}\right) = \frac{1}{N}(\mathbf{R} + (Z-1)\mathbf{T}_{base})$$

$$\cong \mathbf{T}_{base}$$

with a slight overall increase in the real component, and notably retaining all of the ZD's.

Let's now consider the general case where ZD's are indicated as a separate portion (and assume Z ZD's):

$$\mathbf{T}_{base} = \left(\mathbf{R} + \mathbf{T}_{imag}\right) + \sum_{i \in all} a_i\mathbf{T}_{ZD,i}$$

and

$$\text{Em}(\mathbf{T}_{base}) = \frac{1}{N}\left(\left(\mathbf{R} + \mathbf{T}_{imag}\right) + (Z-1)\mathbf{T}_{base}\right) \cong \mathbf{T}_{base}$$

with a slight overall increase in the non-ZD part while still notably retaining all of the ZD's. There is thus conservation of ZD's, suggesting association of ZD's with matter/energy and the conservation of the latter seen in the emanated propagator formalism. The nature of this matter association is still unclear, however, until we consider the next condition on the emanator.

Let's now consider the form of the emanator when it is summed into the 4 chiralities (with 78 or 72 card decks dependent on form):

$$\text{Emanation}(\mathbf{T}_{base}) = \frac{1}{N}\sum_{K \in 4\ chiralities}\mathbf{T}_{base} \bullet \overline{T}^{(K)}_{chiral}$$

and, thus

$$\mathbf{T}_{base} \bullet \overline{T}^{(K)}_{chiral} \neq 0$$

In this context the zero divisors in the base force an unexpected constraint if we require that no elimination of chirality (thus violation of emanator achirality) can occur. In other words, we hypothesize the emanation is constrained such that it is analytic on the log of the products such that zero's are eliminated from the maximal analytic domain.

On the other hand, suppose the form of the emanator can be written as a sum on achiral groups. Such groups *can* be zeroed-out, which describes a form of wave-collapse or measurement filter for the theory:

$$\text{Emanation}(\mathbf{T}_{base}) = \frac{1}{N}\sum_{K \in achiral\ group}\mathbf{T}_{base} \bullet \overline{T}^{(K)}_{achiral}$$

and, thus, we can have:

64

$$\mathbf{T}_{base} \bullet \overline{T}^{(K)}_{achiral} = 0.$$

From the preceding results we then see that we can formulate a hypothesis for the meromorphic precipitation of quantum matter with dimensionful action, where:

(1) The trigintaduonion emanator is doubly analytic, where the first analyticity is in regards to removing the zero-divisors from the domain of the trigintaduonion emanator by means of analytic operations to remove the zero-event for each of the effective dimensions, giving rise to a dimensionless 'action' S^* and a quantum of that action given by:

$$|h^*| = \left(\frac{1}{2\pi m}\right)^{29^*}, where\ m = 2.$$

While the second analyticity is in regards to the resulting sum on associated zero-divisor paths. Upon analytic operation (Wick rotation) we arrive at a sum on paths whose phase is given by a dimensionful action with respect to a dimensionful quantum of action (Planck's constant):

$$S^*/h^* \rightarrow S/h$$

(2) We arrive at large-parameter integral over paths, that is highly oscillatory given $|h^*| \approx |h| \ll \propto < 1$, and it must satisfy the quantum deFinetti relation [19], to give rise to a real action, with:

$$S = \int Ldt$$

where the real-valued Lagrangian is selected to be consistent with the standard model (plus small extension) indicated in the prior section.

Thus, the dimensionless quantum arises from analyticity in the form of a meromorphic function association to each of the 29D in a given chiral propagation, where associated zero divisor (ZD) surgery gives h*<<α<1 since each ZD has 29 real component dimensions (plus a remnant of imaginary dimensionality, thus effective dim=29* [37]), and where a point-like location is given by the location of the cut-out.

The dimensionful quantum arises from Wick rotation from real to pure imaginary (with ZD cut-outs) such that (S*/h*) with discrete time steps 'n' Wick rotates to S/h with dimensionful time 't'. The exact numerical relation h* →|h| may be a truly random emergence that will never be defined further (simply part of the 22 parameters). The main constraint, which is satisfied, is that the quantum be very small, giving rise to an oscillatory integral formalism. A shift in the small constant can't be explained further with the current development of the theory.

5.2.4 Maximal emanator analyticity via removal of zeros-divisors (using split-T form)

ZD-removal was originally done on a hypothesized $[\mathbb{R}\otimes]^{29}$ decomposition but this decomposition is now accomplished without approximation, by construction, by use of the split-Cayley form. The $[\mathbb{R}\otimes]^{29}$ is then taken to be real part of a $[\mathbb{C}\otimes]^{29}$ analytic continuation, where a double-pole residue [22] leads to a derivation for a '1/big' QM-factor to be discussed later. The decomposition (via split) then provides a complex extendable noise propagation effective dimensionality 29*, thus have derivation of 'h*' in [22]. The presumption was that ZD-encounters could occur in the (max) perturbed 10D chiral-T propagation since it is probing the 32D T-space where ZD's are known to exist. With the split-formulation, there are ZD encounters for the 10D chiral-T propagation even with no perturbation. This will be more suitable for the ZD surgery argument since now instead of having to be infinitesimally near a chiral-T trajectory, we can simply take the chiral-T trajectory.

5.3 MIE Recap
5.3.1 Maximum Information Emanation: Fiat Numero
(1) Maximal Information flow without perturbation is in 10D chiral subspace of 32D trigintaduonions [1];
(2) Maximal perturbation is by amount $\alpha=1/137^*$ into the enveloping 32D space [6];
(3) Maximal Information flow with perturbation is 32D, where all chiral subtypes are summed and normalized, with the four 29*D chiral propagations possible in the analytically continued 32D complexified space (64D). We thereby arrive at the edge-of-chaos maximal information emanation relation, where $\alpha^{-1} = (c_\infty)^\gamma$, with its suggestion that the universal evolution is at a fractal boundary, and thus fractal itself [37].

5.3.2 Maximum Information Emergence: Logos Incarnate
(4) The dimensionless quantum arises from analyticity in the form of a meromorphic function association to each of the 29D in a given chiral propagation, where associated zero divisor (ZD) surgery gives $h^*<<\alpha<1$ since each ZD has 29 real component dimensions (plus a remnant of imaginary dimensionality, thus effective dim=29* [37]), and where a point-like location is given by the location of the cut-out.

66

(5) In what follows, we shift from emanator projection to discrete-time propagation with (S*/h*) and, most notably, a shift from propagation in terms of trigintaduonion emanation steps comprising trigintaduonion multiplications to the more conventional propagation in terms of complex propagators comprising multiplication of complex functions of a complex variable. The shift from 32D emanator numbers to 2D propagator complex functions is necessitated by consistency with the maximal info flow hypothesis and the known constraints of the quantum deFinetti relation to information flow with complex propagators [19].

(6) The dimensionful quantum arises from Wick rotation from real to pure imaginary (with ZD cut-outs) such that (S*/h*) with discrete time steps 'n' Wick rotates to S/h with dimensionful time 't'. The exact numerical relation h* → |h| may be a truly random emergence that will never be defined further. The main constraint, which is satisfied, is that the quantum be very small, giving rise to an oscillatory integral formalism. A shift in the small constant can't be explained further with the current development of the theory. Experimental data is used to justify the dimensionful choices of time in seconds, etc.

5.3.3 Complex functions, mappings, and fractals (2D → 2D transforms)

The 2D plane can be mapped repeatedly to itself. For such situations the asymptotic behavior can be examined. Take for example the classic Mandelbrot set mapping where

$$z_{new} = f(z_{old}) = (z_{old})^2 + c$$

For the Mandelbrot set the stability boundary has fractal dimension 2 [38]. One might guess a number $1 \leq frac \leq 2$, but to actually reach the dimension 2 shows an optimality for the Mandelbrot set in this regard that will be called upon later where it will lead to a "zero-divisor" occurrence under that circumstance, thereby effecting a second order zero. The order of the zero will be relevant to the initial dimensionless Planck "h*" calculation (when using the residue theorem from complex analysis).

5.3.4 General Zero Divisors in Cayley algebras

The division algebras do not have zero divisors and comprise the first four algebras of the Cayley family: the real numbers, the complex numbers, the quaternions, and the octonions. Beyond octonions the

Cayley algebras have zero divisors. The next two Cayley algebras are the sedenions and the trigintaduonions. Let's begin by analyzing the zero divisors for the sedenions. Consider the situation:

$$S_1 \cdot S_2 = 0, where\ S_1 = (O_{1L}, O_{1R}) \neq 0\ and\ S_2 = (O_{2L}, O_{2R}) \neq 0$$

S_1 and S_2 in the above form a zero-divisor pair. Let's carry the analysis to the level of octonions to extract more manageable relations:

$$(O_{1L}, O_{1R}) \cdot (O_{2L}, O_{2R})$$
$$= ([O_{1L} \cdot O_{2L} - O_{2R}^* \cdot O_{1R}], [O_{1R} \cdot O_{2L}^* + O_{2R} \cdot O_{1L}]).$$

For the last expression to be zero, it must be zero component-wise, and we arrive at two relations:

$$O_{1L} \cdot O_{2L} = O_{2R}^* \cdot O_{1R} \quad and \quad O_{1R} \cdot O_{2L}^* = -O_{2R} \cdot O_{1L}$$

Let's consider the simplest case, where the four octonions are unit octonions

$$\{O_{1L}, O_{2L}, O_{1R}, O_{2R}\} \in \{e_i\}\ i = 0..7,$$

where

$$e_i \cdot e_j = \begin{cases} e_j\ if\ i = 0 \\ e_i\ if\ j = 0 \\ -\delta_{ij}e_0 + \varepsilon_{ijk}e_k \end{cases}$$

where the antisymmetric tensor is 1 when $ijk =$ {123,145,176,246,257,347,365}. The conditions are then simplified to (unit octonion for O_{1L} is labeled as e_{1L}):

$$\varepsilon_{(1L)(2L)(k)} = \varepsilon_{(1R)(2R)(k)} \quad and\ \varepsilon_{(1R)(2L)(k)} = -\varepsilon_{(1L)(2R)(k)}.$$

Consider the first zero divisor indicated, where:

$$\varepsilon_{(1)(4)(5)} = \varepsilon_{(3)(6)(5)} \rightarrow \{1L = 1, 2L = 4, 1R = 3, 2R = 6\}$$

Thus, $S_1 \cdot S_2 = 0$, with $\{S_1, S_2\}$ is a zero divisor pair, if $S_1 = (e_1, e_3)$ and $S_2 = (e_4, e_6)$. If converted to sedenions (see multiplication table at [48]):

$$S_1 = (e_1, e_3) = (\hat{e}_1 + \hat{e}_{3+8}) = (\hat{e}_1 + \hat{e}_{11})$$
$$S_2 = (e_4, e_6) = (\hat{e}_4 + \hat{e}_{6+8}) = (\hat{e}_4 + \hat{e}_{14})$$
$$S_1 \cdot S_2 = (\hat{e}_1 + \hat{e}_{11}) \cdot (\hat{e}_4 + \hat{e}_{14}) = \hat{e}_5 + \hat{e}_{15} - \hat{e}_{15} - \hat{e}_5 = 0.$$

So we see that there are zero divisors in the sedenions, with a concrete example above, for any set of indices that can be chosen for the antisymmetric tensor in its first two positions. Thus i can take 7 values and j 6 in the relation: $\varepsilon_{ij(k)}$, so 42 cases. A similar set of relations exist for the negative antisymmetric tensor indices for another 42 cases. So there are 84 such zero divisors.

There are only 84 discrete instances of zero divisors for the sedenions. Can this number be increased by relaxing assumptions in our derivation above? (1) Can we interpolate with $S_1 = (\tau e_1, (1 - \tau)e_3)$ for some variety of τ? If we try this the antisymmetric tensor forces the single (equilibrium) case where $\tau = 1/2$. (Norm=1 would force this as well.) (2) Can we generalize solutions of the form $(\hat{e}_1 + \hat{e}_{11}) \cdot (\hat{e}_4 + \hat{e}_{14}) = 0$ to sedenions consisting of more than the addition of two unit sedenions? (In turn, this traces back to assuming the octonion decomposition consisted of unit octonions.) For this to work we would have an expression:

$$\{3 \; \hat{e}_i \; term\} \times \{3 \; \hat{e}_j \; term\} = \{9 \; \hat{e}_j \; term\}$$
$$\rightarrow odd \; terms, can't \; cancel \; pairwise, no \; zd's$$

trying again with expression with four unit vector terms:
$$\{4 \; \hat{e}_i \; term\} \times \{4 \; \hat{e}_j \; term\} = \{16 \; \hat{e}_j \; term\} \rightarrow need \; 4 + 4 + 8$$
$$= 16 \; indep. \; imag's$$

For the latter case, we see that to have an expression with a sedenion comprising 4 unit sedenions multiplied by another such, the resulting expression will have 16 product terms, for which pairwise cancellation is possible (16/2=8 new imaginaries introduced), so 4+4+8=16 independent imaginary components are needed (if we have enough imaginary terms to accommodate). This is not the case for sedenions but is the case for trigintaduonions. This exhausts the possibilities for sedenions, thus there are only the 84 sedenion zd's indicated. Continuing this analysis for trigintaduonions:

$$\{5 \; \hat{e}_i \; term\} \times \{5 \; \hat{e}_j \; term\} = \{25 \; \hat{e}_j \; term\}$$
$$\rightarrow odd \; terms, can't \; cancel \; pairwise, no \; zd's$$
$$\{6 \; \hat{e}_i \; term\} \times \{6 \; \hat{e}_j \; term\} = \{36 \; \hat{e}_j \; term\} \rightarrow need \; 6 + 6 + 18$$
$$= 30 \; indep. \; imag's$$

The latter case is still possible for trigintaduonions, which have 31 imaginary components.

Suppose for trigintaduonions instead of the antisymmetric tensor we have some more general third-rank tensor with no indices the same, again we will have a relation with the first two indices (ranging over sedenions), with two sign forms, giving rise to no more than 15*14*2=420 independent discrete trigintaduonion zero divisors. If we view the approximate number to be four times that of the sedenions since 1 T-multiplication can be turned into 4 S-multiplications we could argue for 4x84=336 trigintaduonion zero divisors. Regardless, it is a discrete set as with the Sedenions.

5.4 ZD-removal and Integrals with large parameter
A review of oscillatory integrals now follows. This mathematics is critical to the path-integral quantization program. It traces to Laplace's method of steepest descents, then to the work of Stokes and Lord Kelvin, then to the work of Erdelyi and others [49-51], before its incorporation by Feynman into his path integral formulation of QM and QFT, where the most precisely tested result in physics was then shown with QED [52-54].

So far we've seen sums involving possibly an infinite number of products during an emanation step. These are related to a definite integral with appropriate measure. Thus we arrive at a discussion of definite integrals. Now if the integrand has maxima or stationary points the definite integral is often dominated by those regions (to be shown momentarily), so the focus turns to an asymptotic analysis of the integrals about those internal points (and boundary points), e.g, an asymptotic expansion analysis. (The easiest asymptotic expansion for a definite integral is obtained by repeated integration by parts.)

For what follows we are interested in the definite integral with large parameter:

$$f(x) = \int_a^b e^{xh(t)} dt$$

where x is very large (or grows large). Under these circumstances we are interested in any critical point t_0, where $h'(t_0) = 0$, and we write $h(t)$ in terms of its Taylor expansion about that critical point:

$$f(x) \approx e^{xh(t_0)} \int_a^b e^{-\frac{1}{2}x|h''(t_0)|(t-t_0)^2} dt$$

70

and we then approximate with the integration bounds at infinity to get the standard Gaussian integral. Thus, we get the asymptotic expansion solution for large x:

$$f(x) \approx e^{xh(t_0)} \sqrt{\frac{2\pi}{x|h''(t_0)|}} \quad for\ large\ x.$$

Let's now consider the more general form originally studied by Laplace:

$$f(x) = \int_a^b g(t)e^{xh(t)}dt$$

similar analysis can proceed if we make the substitution $h(a) - h(t) = u^2$ (due to Laplace), where a factor of $\frac{1}{2}$ is introduced since the eventual domain of integration will be $\{0,\infty\}$ not $\{-\infty, \infty\}$:

$$f(x) \approx g(t_0)e^{xh(t_0)} \sqrt{\frac{-\pi}{2xh''(t_0)}} \quad for\ large\ x.$$

If we consider the last integral but generalized to x a large complex variable, and g and h to be analytic functions of complex t (first studied by Riemann and Debye [51]), we have similar analysis but where we start by deforming the path of integration as much as possible to coincide with paths of steepest descent. What we seek, however, is not the full complex generalization, but the alternate form that might be arrived at by analytic continuation from the initial (pure) real form to a pure imaginary form (via a Wick rotation). For this we arrive at the integral:

$$f(x) = \int_a^b g(t)e^{ixh(t)}dt$$

where x is large and positive and $h(t)$ is real as before (but with the factor of i, now effectively pure imaginary). To solve this type of integral the integral is dominated by terms not cancelled, i.e., where the phase is stationary in the integration. This occurs at the critical points (and end-points) as before, but represents a slower convergence or domination about the critical point than that in the exponential fall-off case dealt with by Laplace. The method of stationary phase was initially developed by Stokes and Kelvin [49]. By similar arguments to that shown above we then arrive at the solution:

71

$$f(x) \approx g(t_0)e^{[ixh(t_0)+\frac{i\pi}{4}]}\sqrt{\frac{2\pi}{xh''(t_0)}} \qquad for\ large\ x.$$

In the result above, the integral is dominated by the region around the stationary point. Since this integral is written $\{-\varepsilon, +\varepsilon\}$, which is extended to $\{-\infty, \infty\}$, we get the standard Gaussian integral factor. If we generalize to g and h analytic, then the point of stationary phase is a saddle-point in the complex plane and using methods like in the method of steepest descent, the integration path is deformed to traverse the stationary-phase saddle-points (from stirrup to stirrup) such that the same result above is obtained as the dominant contribution as x grows large [51].

5.4.1 Maximal emanator analyticity via removal of zeros [46]
The sum over the zero-divisors means that the part of the emanator requiring analytic 'repair is given by:

$$\sum_{\{S^*_i\}} \mathbf{T} \bullet \bar{T}_{chiral}(S^*_i) \rightarrow \sum_{\{S^*_i\}} e^{\mathbf{T} \bullet \bar{T}_{chiral}(S^*_i)} \rightarrow \sum_{\{i\}} e^{S^*_i/h^*} .$$

Where use is made of the fact that approach to zero-divisor (ZD) is purely involving imaginary components. The shift to exponential form will be explained with the choice of analytic continuation or 'repair' described in the next section. The sum on ZD events (for all 'time') can thus be described as a sum on (ZD) paths. We can now see the identification of matter with the zero-divisor 'residues' that occur when imposing maximal analyticity. The dimensionality of possible ZD's (for trigintaduonions) thus indicates a dimensionality on "paths," with result:

$$\sum_{zd's} e^{S^*_i/h^*} \rightarrow \int_{zd\ paths} e^{S^*(i)/h^*} .$$

Now do a Wick rotation and go from real dimensionless iteration-number to imaginary dimensionful action, with dimensionful Planck's constant. We then get the highly oscillatory integral that is the basis of quantum field theory and quantum mechanics, with their classical and semiclassical reductions. With matter reified by Wick rotation, we go from an integral on zd paths with large parameter $1/h^*$ to an integral on matter paths with large parameter $1/h$. We, thus, maintain the large-parameter form as we go from a Laplace-type integral to a Stokes-type integral, and thus arrive at a path integral formulation:

$$\int_{zd\ paths} e^{S^*(i)/h^*} \rightarrow \int_{matter\ paths} e^{iS(i)/h}, \quad where\ S(i) = \int L dt.$$

In what follows, the shift from emanator projection to discrete-time propagation with (S*/h*) and, most notably, a shift from propagation in terms of trigintaduonion emanation steps comprising trigintaduonion multiplications to the more conventional propagation in terms of complex propagators comprising multiplication of complex functions of a complex variable. The shift from 32D non-associative emanator numbers to 2D (complex, associative) propagator functions necessitated by consistency with the maximal info flow hypothesis and the known constraints of the quantum deFinetti relation to information flow with complex propagators [46].

5.4.2 Zero-divisor removal at component level [46]
In what follows we will require zero removal for analyticity on the log of the trigintaduonion products for a particular chirality of emanation. Let's now calculate the zero removal residue seen as a product of each of the 29* effective dimensions of the analytically-continued real components. Recall that:

$$\oint_C \frac{1}{z} dz = \oint_C d(\ln z) = 2\pi i \quad (simple\ pole).$$

where C is a contour that encloses the pole, which generalizes to:

$$\oint_C d(\ln f(z)) = \sum_{zeros} 2\pi m i \quad (multiple\ zeros),$$

where f has multiple zeros of order m, and where the last result requires that $f(z)$ be analytic throughout the domain, D, with boundary C inside that analytic domain (and D is simply connected). Let S^*_i be the zeros of $f(z)$ where at lowest order $f(z)$ has a double zero at each of the S^*_i according to the maximum fractal dimension possible for the boundary condition at the edge-of-chaos (where the dim=2 boundary dimension is actually the case for the Mandelbrot Set [38]). Let's use this information to parameterize the approach to the zeros:

73

$$\mathbf{T} \bullet \bar{T}_{chiral}(z) \propto \prod_{d=29*} (z - S^*{}_i)^2,$$

thus, for multiple zeros:

$$\oint_C \frac{d}{dz}(\ln[\mathbf{T} \bullet \bar{T}_{chiral}(z)])dz = \sum_{zeros\ 29*} \prod 4\pi i.$$

Focusing on just one of the zeroes and the line integral dominated by a local, stationary phase, contribution, we need to integrate and set $z = S^*{}_i$:

$$d(\ln[\mathbf{T} \bullet \bar{T}_{chiral}(z)]) = 4\pi i^{29*} dz.$$

and, with choice of integration constants (phase factors):

$$\mathbf{T} \bullet \bar{T}_{chiral}(S^*{}_i) = e^{S^*{}_i/|h^*|}.$$

Summing on the zeros of the latter expression:

$$\sum_{zeros\ S^*{}_i} \mathbf{T} \bullet \bar{T}_{chiral}(S^*{}_i) = \sum_{zeros\ S^*{}_i} e^{S^*{}_i/|h^*|}.$$

Thus, the general form of maximal, analytic, information emanation gives rise to a sum on residue-like terms associated with each of the zero-divisors (zd's), and an 'action' variable is indicated to result from the parameterization of the approach to each of the zero-divisors, with their individual actions additive (phase contributions multiplicative) for parts contributing to a particular zd. The sum over all the zd's will, upon analytic continuation, be associated with a sum over paths. The zd action variable is written in the form of the integral of a functional along a path parameterized by 'time', with the usual definition for Action if the functional is the Lagrangian:

$$\sum_{zeros\ S^*{}_i} e^{S^*{}_i/|h^*|} \rightarrow \int_{matter\ paths} e^{iS(i)/h}, where\ S(i) = \int Ldt,$$

where the definition for action above is kept to the simple form for a point particle trajectory. More complex forms can be written for field descriptions, where we generalize from point particle forms in various ways, but still with point-like coupling terms. Further generalization to actions describing 1-D objects, string not points, or beyond (n-D objects, or branes) and their trajectories is possible at this point but note how the chain of associations is altered, if not broken. Tracing back to a fundamental issue of analyticity when going from emanator form to propagator form, we saw that analyticity requires *isolated* zeros to not make the entire solution trivially zero. Thus, the fundamental meromorphic 'precipitate' for matter is point and point-based field constructs as we've developed them already. Note that in going from

$$\frac{S^*_i}{|h^*|} \rightarrow i\frac{S(i)}{h}$$

(1) Both ratios are dimensionless, but the quantities on the left are a ratio of pure numbers themselves dimensionless, while the RHS has a ratio of S, the action, with action dimensions, and h, Planck's constant, also with action unit dimensions.

(2) Both $1/|h^*|$ and "$1/|h|$", where the abs value operation on the latter simply drops the dimensionful units, are extremely large numbers and occurring in a phase argument. This *sets up a highly oscillatory integral* (see Appendix in [22]) such that the classical solution $\delta S = 0$ results (if a classical solution exists for the system studied), among other things.

If the emanator is to remain an achiral mix, as well as analytic, then we can't allow the zero divisor events that would drop a chirality mentioned above, where these occurrences are treated as isolated singular events removable from the domain of definition of the emanator by repeated application of the analytic domain 'surgery' (repeated on both events, and for given event, it's different independent components). This analytic 'surgery' occurs for each of the independent component dimensions for a given chiral emanation, and for each of those dimensions it returns a zero-removal 'residue' of $4\pi i$ (with an extra factor of 2 since a double zero at the fractal boundary). We found in [37] that the effective dimension is 29*, thus the remnant of the surgery for each zd removed is:

$$\frac{1}{|h^*|} = (4\pi)^{29^*}.$$

So far we've examined the zd's and their residues on the $(C \times)^{\wedge}29^*$ local factorization. If we consider the complex structure in a global sense, any meromorphic function on a sphere, due to its compactness, must be rational. Evaluation of zero divisors happens to occur when we consider the product of two pure unit-norm imaginary trigintaduonions, thus on a 31-sphere. A quantization on the matter has occurred in the sense that the meromorphic function must associate with *isolated* singularities, so a discrete, countable, number of matter-associated events must occur. Also note that we have:

$$|h^*| = 6.630 \times 10^{-33}$$

$$vs$$

$$Plank's\ constant\ \ h = 6.626070 \times 10^{-34} J\ s$$

where we only need these two 'h' numbers to satisfy the same extreme smallness property in order to obtain integrals with large parameter and thus *a highly oscillatory integral with stationary phase domination.*

5.5 Achieving Dimensionality (e.g., physical units)

The next issue is how to dress up the key parameters with dimensionful units, to arrive at the standard physics formulations, when the original formalism is purely algebraic, albeit with dimensionless constants already found to exist (α). This will be accomplished by Wick rotation from integration on real terms to integration on imaginary phase contributions. In making this analytic continuation we introduce units via transforming a dimensionless ratio of dimensionless numbers to a dimensionless ratio of *dimensionful* numbers. We also go from summing on zero-divisor associated terms to summing on zero-divisor associated 'paths'. The summations on path add according to their phase, the latter dependent on the action expressed as

$$S = \int L dt$$

where time emerges as the parameterization of the path and the Lagrangian is that indicated by the Standard Model (plus Hilbert General Relativity term to be discussed further in the Discussion). Analyticity on this integral (and all integrals encountered thus far), in the form of the Wick rotation especially, is encompassed by the description of complex time Euclideanization in later sections. Note, this describes a doubly analytic structure (at level of emanator and at level of propagator). Since the Wick rotation on the trigintaduonion (32D) objects represents use of an analytic complex structure to extend each of the real components to complex components, we have an analytic extension off of the 32D Cayley algebra into the enveloping 64D Cayley algebra. This is best seen in terms of the well-defined exponential map described earlier where T → C × T by means of exp(iT)=cosθ+iTsinθ. Another way to view the analytic structure is not as an added structure but the residual structure of the hypercomplex selection for maximal transmission settling to the highest order Cayley 'sub-algebra' it can manage. As such, the emanation process that arrived at the 32D Cayley algebra did so in a context where an analytic 64D Cayley algebra extension already existed (albeit briefly).

For a zero-divisor to occur with the T_{base} the real component must be zero, but this is possible for the base trigintaduonion in the Emanator. The number of emanation steps to a zero-crossing event, with random-walk statistics on the real component, is revealed in [37],and will be analyzed in Ch. 10.

Chapter 6. Why 22?

6.1 Why 22 parameters?

The Trigintaduonion emanation hypothesis strongly indicated the possibility of 22 'fundamental' parameters in early efforts [15]. Here we see a clearer mathematical argument for this being the case. Consider the emanator for the excluded case of a pure real common octonion δ:

$$T_{chiral}^{(k)} = \begin{cases} ((\delta,\alpha),\beta) \\ ((\alpha,\delta),\beta) \\ (\beta,(\delta,\alpha)) \\ (\beta,(\alpha,\delta)) \end{cases}.$$

We get no sub-octonionic mixing for this type of emanation: $T_{em,} = O \times H$, where $O = 1$. Such an emanation, acting on T_{base}, in order to remain achiral, will change nothing vis-a-vis future achiral T_{em} products, thus it is associated with a "constant of the emanation". How many such emanations, effecting no change, are there? They correspond to the non-octonion based fluctuations in the evaluation of the 78 deck described previously, of which there are 22.

6.2 Computational examination of the 22 parameters in Chiral Emanation

In the tables that follow are shown the emergent parameters when alpha-perturbations (perturbations with max perturbation alpha) are injected for each of the 22 non-propagating parameters. Regardless of injection parameter, if perturbation exceeds alpha, the norm=1 relation fails, and propagation eventually dies with norm ≈ 0. This is to be expected given the identification of alpha as the max-perturbation limit in [6]. What is odd is that if perturbation is less than alpha, but still in the vicinity of alpha, norm=1 behavior appears to eventually fail (after millions of iterations, and using bignum precision) as can be seen in the real component eventually falling to zero. In other words, the iterative procedure underlying the propagator definition, not surprisingly, is giving rise to fractal behavior (and abrupt transitions). I say not surprisingly because the single parameter noise injection that we are using (in repeated multiplicative iterations) is such that we've set up an iterative process with a 1-dim parameter space and are seeing possible fractal behavior -- a well-known phenomenon in 1-D complex systems.

Thus, the results that follow are *preliminary estimates* on the 'letters of reality' (actually numbers in this numerogenesis algebraic theory) in that both the noise injection method can lead to artifacts, and due to the slow process of doing bi-sedenion multiplication with bignum(50).
In the experiments tabulated in what follows we consider unit norm chiral bi-sedenion propagation. In particular, we consider unit element chiral bisedenion propagations, with alpha perturbations introduced, separately, at each of the non-propagating bi-sedenion parameters. If working with a perturbation greater than alpha we expect the real component to start at one (we begin with a unit element chiral bisedenion) and eventually decay to zero, and cross-over to negative values, as it begins to randomly walk. To a lesser extent, and with fractal structure, this also appears to be true for perturbations introduced that are in the vicinity of alpha but less than alpha. For perturbations precisely at 'alpha', we expect the real component to decay/search for a while, but to then asymptote/lock-on to a particular non-zero (emergent) value with well-defined variance about that asymptote, and never crossing zero. In other words, if we want to propagate one bit of information via the real component asymptote of the bi-sedenion indefinitely, it is hypothesized that we can do so using alpha perturbation propagators. In Fig. 1 is shown the Histogram on real components (rc) observations after each multiplicative iteration, where an emergent rc=0.971 appears in the first 60,000 propagation iterations.

To recap, for 'off-shell' bi-sedenion propagation at maximum perturbation amplitude alpha, we examine the behavior of the real component (rc) of the bi-sedenion. This is because we are effectively describing propagation starting with a unit bi-sedenion (so only have rc=1 nonzero), followed by multiplicative propagation steps by way of bisedenions perturbed by at most the fraction alpha into the bisedenions 31 imaginary components. We consider each of the 22 possible non-propagating parameters in separate perturbation-at-alpha analyses, where the emergent behavior on the rc value is obtained. The Results in Table 1 show the emergent 22 parameters when propagation is done with precisely alpha perturbation, where alpha is taken to be the highest precision value known provided by QED (1/137.035999070) [55] (which appears via $\delta=0.001459470514006$ in the code).

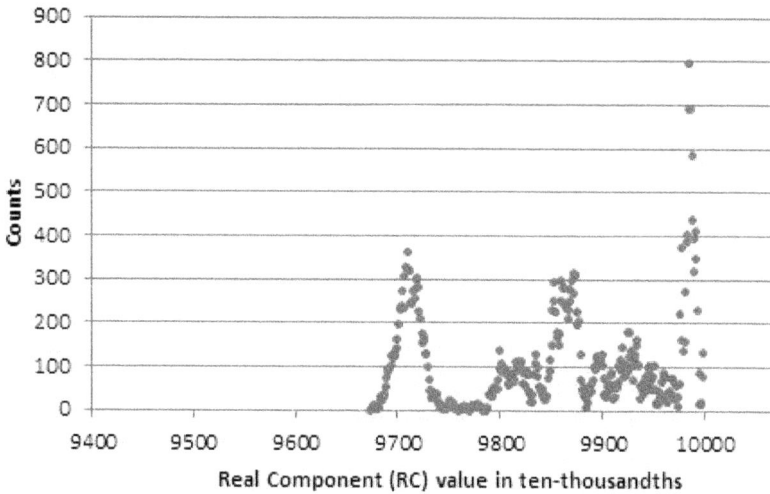

Fig. 1. Histogram of real-component values observed in the first 60,000 iterations of alpha propagation, where the perturbation parameter is δx_{30}. As the propagation begins the RC value is at 1.0=10,000/10,000, so at the rightmost dot in the histogram. As the multiplicative operations proceed, the rc value decays thru the range to rc =0.9750, then begins to catch the asymptote with mean at 0.9710.

The alpha propagations are examined for each of the 22 different non-propagating components, with each taken individually as sole source of non-propagating perturbation in its respective alpha propagation experiment. Since this process is selecting propagators somewhat arbitrarily, perhaps not as much utility can be extracted from the asymptotic RC values as from their *variance* information. In other words, by injecting noise perturbations into each of the 22 parameters separately (like playing a recorder with only one hole depressed at a time), computational experiments are attempting to arrive at information respective to 22 parameters, but may do so in a mixed form not so useful when expressed in the RC values. Furthermore, the Gaussian distributions that appear to be emergent at the asymptotes have variance values (or their inverses as shown in Table 1) that may provide the most utility. In essence the variance can be thought of as describing a statistical restoring force that's occurring in the bi-sedenion propagation due to the odd properties of bisedenion in general, e.g., they are: non-associative, non-commutative, have zero-divisors, and lack of inverse due to lack of norm. Bisedenion properties are not theoretically fully understood at this time,

81

thus the computational efforts described here (and in [1,6]) to try to resolve matters further.

Off-shell parameter	Asymptotic Real Component (RC)	Asymptotic RC FWHM	Asympt. RC 1/Variance
δx_9	0.9823	0.0047	*246,819*
δx_{10}	0.9361	0.0044	*281,623*
δx_{11}	0.9585	0.0030	*605,803*
δx_{12}	0.9856	0.0021	*1,236,332*
δx_{13}	0.9953	0.0017	*1,886,583*
δx_{14}	0.9343	0.0029	*648,302*
δx_{15}	0.9745	0.0023	*1,030,666*
δx_{17}	0.9644	0.0039	*358,463*
δx_{18}	0.9745	0.0050	*218,089*
δx_{19}	0.9799	0.0060	*151,450*
δx_{20}	0.9792	0.0053	*194,098*
δx_{21}	0.9639	0.0048	*236,641*
δx_{22}	0.9797	0.0028	*695,436*
δx_{23}	0.9593	0.0037	*398,263*
δx_{24}	0.9826	0.0066	*125,165*
δx_{25}	0.9979	0.0012	*3,786,267*
δx_{26}	0.9615	0.0059	*156,628*
δx_{27}	0.9892	0.0041	*324,344*
δx_{28}	0.9497	0.0051	*209,620*
δx_{29}	0.9326	0.0052	*201,635*
δx_{30}	0.9710	0.0022	*1,126,493*
δx_{31}	0.9706	0.0020	*1,363,056*

Table 1. The 22 letters of reality. The 'letters' are emergent real parameters (i.e., just numbers, the 'best' set shown in bold in right column) from an iterative process involving repeated chiral bi-sedenion multiplication. If noise injection at non-propagating ("off-shell") parameter x_9 is introduced then have non-zero components $\{\Delta x_0, \delta x_1, \delta x_2,\ldots\delta x_7, \delta x_8, \mathbf{\delta x_9}, \delta x_{16}\}$. The table lists the off-shell parameter, its asymptotic rc value, the full-width at half maximum (FWHM) of the peak (FWHM=2.335σ), and the inverse of the variance (taken as the best set of 'letters' available at this time).

6.3 Standard Model with SU(2)$_R$ is consistent with 22 parameters
The emanator sum is achiral, but is composed of a sum over chiral T emanators. This collection of chiral emanators, if seeded with a common octonion, with positive and negative fluctuations in each component, leads to a trigintaduonion emanator that has the form $T_{em} \cong O \times H$. As

82

seen in the "sum on phase" analysis naturally indicated in the zero-divisor curing, we will soon consider the properties of the mathematical object of exp(iT$_{em}$), which then explicitly promotes the theory to be C × H × O. According to Dixon [23], it is then possible to obtain an action on the T$_{base}$ space that is precisely the U(1)×SU(2)×SU(3) form desired. This then leads to light matter with maximal species of particles (thus generations), acted on via U(1)×SU(2)$_L$×SU(3), and dark matter with minimal species (sterile neutrinos) acting only via SU(2)$_R$.

From 22-parameter hypothesis, with maximal info transmission, it is apparent that we will get three generation results for the maximal number of interacting particles in one sector (that we will call the "Light' sector accordingly since it has Light, e.g., U(1)), with the remainder left to the "Dark" sector, with only SU(2)$_R$, the sterile neutrino. The reason for light/dark asymmetry is simple, it allows for the maximal complexity of information transmission. Suppose the number of particles in the Light sector is L and that in the Dark is D, then the number of binary interactions is L^2+D^2. Given L+D=constant=C, we find the maxima to occur at {L$_{min}$. R$_{max}$} or {L$_{max}$, R$_{min}$}. The convention is adopted that what we call "Light" matter is the matter that is most interactive, thus we have the labeling {L$_{max}$, R$_{min}$}.

Trigintaduonion emanation theory indicates 22 free parameters with maximum perturbation amount α in the larger 32D trigintaduonion algebra. In the analysis of the possible emanators *analyticity* is indicated in numerous ways, such that this is a core hypothesis for the maximal information propagating solution. This, in turn, indicates analytic surgery via the residue theorem, on the log of the emanator, to create a maximal analytic region. When we Wick rotate from *S**/*h** → *S*/*h*, there should be 22 independent parameters in the action *S* [15], with Planck's constant counted separately. Can we fit the parameters of the Standard Model, with a possible extension for the dark matter mentioned (e.g., sterile neutrinos), and the gravitational constant G, all into that 22 count? Yes, if we adopt the Koide relation [56]. Let's show this by first listing the 19 parameters in the Standard Model:

    (I) 9 Yukawa coupling constants (masses) for the charged fermions
    (II) 5 constants for Weinberg Angle and the CKM matrix (with three mixing angles and CP-violating phase)

(III) 3 Constants for electromagnetic coupling (α), for strong interaction (g3), and strong CP-violating phase ($\theta_3 \approx 0$).
(IV) 2 Higgs parameters: Mass and Vacuum Expectation

If we allow for the neutrinos to have mass, then we get 3 more masses and another 4 constants for the PMNS matrix (three mixing angles and a CP-violating phase):

(V) Extended model: 7 more constants → We, thus, have 26 parameters.

If we add the constant for Gravitation (G) to have all constants for Std. Model + Gravitation, we now have 27 parameters. Note, however, that the α constant is listed above as the EM coupling constant, but isn't really a separate parameter since it is the same for any emergent chiral trigintaduonion emanation. This is all the more apparent if we go with a listing of 19 independent parameters in terms of the g_1 and g_2 coupling constants which share the following relation with α:

$$\alpha = \frac{1}{4\pi} \frac{g_1{}^2 g_2{}^2}{g_1{}^2 + g_2{}^2}$$

So, we take the separate (double) count on α away from the count to get to 26.

The Koide relation [56] was first observed for the three massive leptons currently known:

$$\frac{m_e + m_\mu + m_\tau}{\left(\sqrt{m_e} + \sqrt{m_\mu} + \sqrt{m_\tau}\right)^2} = \frac{2}{3} \ almost \ exactly.$$

To a lesser extend this relation is satisfied for the quarks as well, particularly for the three most massive, where the value is 0.6695. The problem with a simple application to the quark masses is that they are dependent on energy scale. A theoretical explanation for the Koide relation describes how this relation might exist for the masses of a given generation (or family group) [57]. Assuming this or some other theoretical explanation can show that the three masses of a given generation aren't truly three independent parameters, but two. With this correction on 4 generation of masses (now counting the sterile neutrino generation), we arrive at 26-4=22 free parameters as desired., and the emanator theory thus indicates a nearly complete theory in that the 22 parameters are almost all known.

Note that the fine-structure constant α and Planck's constant have very different trigintaduonion emanation origins and uniqueness: α derives from T-emanation directly, without reference to zero divisors, is dimensionless, and is precisely defined. Also, α is one of the 22 fundamental parameters of the trigintaduonion emanation process. Planck's constant, on the other hand, is not one of the 22, arising from the meromorphic matter description instead. Also, Planck's constant is only specified to have an essentially small quantum to establish an oscillatory integral with $h^* \ll \alpha$, and derives from T-emanation when zero divisors are accounted for by way of maximal analyticity.

The emergence of 22 parameter theories
The maximal information propagation occurs for a 10-dimensional doubly-chiral subspace of the 32-dimensional trigintaduonions. The 22 'fixed' dimensions then appear as 22 parameters that 'imprint' on any gauge theory that may emerge from the 10-dimensional propagation (4-dim for space time, 6-dim for a gauge). So, the propagating theory has 22 emergent parameters. Now consider that 10-dim propagation and allow a small perturbation into all dimensions of propagation (including the 22 dimensions). The maximum magnitude of the overall perturbation allowed for unit-norm right (or left) propagation is $\alpha=(1/137.0359998)$, where 137 independent tri-octonionic 'braids' comprise the flow of information within the doubly-chiral subspace of the trigintaduonions. These same numbers $\{10,22,32,137\}$ appear in a number of ancient numerological systems, could it be that they've also identified this maximal information propagation construct in a text analytics setting?

Counting dimensions involved in maximal propagation gives one set of numbers, while counting the degrees of freedom possible (functionally) given that dimensionality is a different matter. First to recap the dimensionalities arrived at, starting with the Cayley Algebra progression of dimensionalities, and partitions therein, during the process of arriving at maximal information propagation:

$$\infty \rightarrow 2^\infty \rightarrow 2^5 = 32 \text{ (the trigintaduonions)} \rightarrow 32 = 10^* + 22,$$

where we have emergent 22 fixed parameters and chiral propagation in dim $= 10^* = 4^* + 6$, where the asterisk denotes the emergent, specific, spacetime reference within the emergent 10D propagation, e.g., we have 4D Lorentzian spacetime with 6D gauge. The 10D propagation is doubly-chiral in the 32 dimensional space, first at the level of the 16 dimensional sedenion subspace, then on the entire 32 dim. trigintaduonion space. The four chiralities of propagation share the same 22 fixed parameters and the same 4D Lorentzian reference frame, thus a 'full' (bosonic) propagation including all four chiralities would have the shared 22 and 4 dimensionalities (or parameters) but with freedom on 6 gauge dimensions for each of the four chiralities. With the discussion on a full propagation involving all chiralites the focus shifts from a counting on dimensions to a counting on functional degrees of freedom (for describing propagation) in that dimensional framework. Any function of a particular parameter (spacetime or gauge) will have independent forward or backward propagations, thus counting for two degrees of freedom each, together with the fixed 22 parameters this give as the number of degrees of freedom at:

Dimensions: single chiral: $10^* + 22 \rightarrow 4^* + 6 + 22$
Degrees-of-Freedom: full doubly-chiral set: $2(4)(7) + 22 = (4)(14) + 22 = 56 + 22 = 78$

6.4 Maximum information transmission (and storage) using 22 letter encodings

The bi-sedenion optimization results for maximal information transmission [16], may also serve another purpose in identifying a new data compression/encoding scheme. If a 22 symbol alphabet is a fundamental, together with 10 parameter weighting scheme, it may be that most of the benefits of that optimal encoding can be obtained by simply remapping to a 22-symbol encoding scheme followed by one of the usual non-lossy compression methods.

6.4.1 Evolution of man-made written systems arrive at 22 letters

It is worthy of note that the existing gematria word-scoring system, and the 22-letter alphabet it works on, would have had hundreds of years to seek optimization at the hands of scribes painstakingly transcribing messages in difficult media (clay tablet, papyrus, etc.). Gematria was an ancient system for encoding messages used by the Babylonians and subsequently adopted by the Hebrew people while under Babylonian control, and it grew in both cases to have religious/numerological significance (part of the standard religion for Babylonians for 800 hundred years, adopted by some Jewish sects up to what is modern day Kabbalah). The gematria word-scoring scheme can be seen as a recursively stable bi-sedenion optimization or propagation (via its choice of 10 dim and 22 letters, in a 32-dim bisedenion). Just as the ancient Greeks were fascinated with the recursive properties of the golden rectangle, incorporating its 'magic' into the choice of the temple foundations and column spacing, etc., the same can be said for the ancient Babylonians/Hebrews as regards the transcription of their religious texts. It appears these texts are 'gematriafied', thereby incorporating the 'magic' or maximal bisedenion info encoding/transmission into their word foundations.

6.4.2 Evolution of biological written systems arrive at 22 letters
Biological information involves two written systems, one with nucleic acid polymer bases (4 DNA or RNA letters), and one amino acid (AA) polymer based (22 AA letters). The nucleic acid polymer system is more ancient, existing before protein (protein is a 'long' AA polymer, while a peptide is a 'short' AA polymer). This is known as the RNA World Hypothesis [58]. The original nucleic acid language was optimized for information storage more than transmission/functionality, and is still retained for that purpose by biological systems.

Evidently the nucleic acid words written in 4 letters, over vast stretches of evolutionary time, began to have groupings for optimal information transmission (where optimality is selected by evolution). In time, these groupings actually provided a template and mechanism for creating 'words' in a new, amino acid based, alphabet system. These new types of biological words were tasked with exploring the entire space of functional possibilities (of biomolecules), thus lent themselves to the typical 22-letter, 10/11 weight-factor (dimension), optimization result. Another example of the same encoding scheme, just different naming, in ancient gematria the 22 letters are separated into three groups (with different 'weight' factors accordingly): the 3 'mother' letters (aleph, mem, shin); the 7 double letters (beth, gimel, daleth,kaph,pe,resh,tau); and the 12 simple letters (he, vau, zain, cheth, teth, yod, lamed, nun, samekh, ayin, tzaddi, qoph). If we want to look for a similar splitting into subgroups on properties of the amino acids we don't have to look far, as the 22 amino acid 'letters' are similarly split into groups of 3, 7, 12: there are 3 punctuation types of letter (2 stop types, one start); 7 hydrophobic AAs (alanine, isoleucine, leucine, phenylalanine, proline, tryptophan, valine), excluding methionine, which codes for 'start'; and 12 hydrophilic AAs

87

(arginine, asparagine, aspartate, cysteine, glutamate, glutamine, glycine, histidine, lysine, serine, threonine, tyrosine). Other parallels exist as well, and not surprisingly so, if they are both an optimized information transmission/functionality processes that must relate to bi-sedenion propagation (or weighting) in a 10 element scheme, with emergent 'letters' in that scheme numbering 22.

Since this 22 letter AA alphabet is highly suggestive of a successful bi-sedenion encoding optimization, this begs the question of what might biochemically correspond to the 10 (or 11) elements in the gematria-like optimized encoding scheme? This would indicate a 'scoring' on protein words as with text words. Such scorings on hydrophobicity, etc., have already been employed by biologists, but it might be possible that an optimal 'gematria' scoring system is already built-in, waiting to be discovered. So what might make up the elements in weighting the different amino acids? We have the standard hydrophilic vs hydrophobic; polar vs nonpolar; small vs large; monosteric vs allosteric (as a split that sometimes occurs, especially with complexities of hydrogen bonding); and AA's mostly reserved for punctuation (1 start or 2 stops). Counting the polarization pairs of 'this vs that' as two elements impacting weighting, then the number of elements impacting the weighting assigned to a 'letter' (AA in this instance) is 10 or 11 according to whether allostericity is present. So an exact match with what would be expected of an optimized bi-sedenion formulation. It is interesting to note that the richness of the 10/11 parameter polymer scoring/functionalization-capabilities in aqueous solution may not be possible in non-aqueous solutions. Life may require water-based biochemical processes to effectively code for functionality in the geologic time-scales in which it can exist (before catastrophic environmental change).

6.4.3 Propagation of physical information arrive at 'words' consisting of 22 letters

The breakdown on the emergent 22 parameters in the case of the 19 parameters of the standard model plus the minimal neutrino extension [59,60] (have three new mass parameters for three sterile neutrinos) can similarly be done. The minimal extended standard model has 3 generations of four mass parameters, so 12 masses to correspond with the 12. Likewise, there are 7 'angle' or 'coupling' parameters, and 3 other parameters having to do with Higgs or phase.

6.5 Split redux: Effective dimension $29^* = \ln(\alpha^{-1})/\ln(\sqrt{C_\infty})$ and the small-h derivation

6.5.1 The maximal noise propagation dimension hypothesis

$$\alpha^{-1} = \left(\sqrt{C_\infty}\right)^{29^*}, \quad where \; dim. = 29^*.$$

6.5.2 The 'smallness' of Plank's constant explained

We now consider zero-divisor (zd) removal for a hypercomplex function in the 29* dimensional noise propagation space. If the emanator is to remain an achiral mix, as well as analytic, then we can't allow the zero divisor events that would drop a chirality mentioned above, where these occurrences are treated as isolated singular events removable from the domain of definition of the emanator by repeated application of the analytic domain 'surgery' (repeated on both events, and for given event, at component level). This analytic 'surgery' occurs for each of the independent component dimensions for a given chiral emanation, and for each of those dimensions it returns a zero-removal 'residue' of $4\pi i$ (with an extra factor of 2 since a double log zero at the fractal boundary). We found in [37] what the effective dimension is 29*, thus the remnant of the surgery for each zd removed is:

$$\frac{1}{|h^*|} = (4\pi)^{29^*}.$$

In the above, we assumed a $(R \times)^{\wedge}29^*$ local factorization existed and previously considered explicit cases with chiral T's that were infinitesimal variations on unity in order to ensure decoupling of terms (at first order) to effect a $(R \times)^{\wedge}29$ factorization. None of this constraint is needed in the split formulation, where we trivially have the factorization due to the split basis. We then analytically extend to being the real part of a $(C \times)^{\wedge}29$ that would then be extended to the maximal (fractal) dimension 29*: thus $(C \times)^{\wedge}29^*$. We have:

$$|h^*| = 6.630 \times 10^{-33}$$

compare this with Planck's constant $h = 6.626070 \times 10^{-34} J\,s$, where the two 'h' numbers satisfy the extreme smallness property, thus providing integrals with large parameter and thus *a highly oscillatory integral with stationary phase domination, the foundation of the quantum mechanics path integral formulation.*

Chapter 7. Lorentz Invariance

A paper published in 1917 describes Lorentz invariance for complex bi-quaternions [61] (modern notation [62,63]). The chiral bi-sedenion emanator described in [1] is re-examined here with its general octonion, which can be either a regular Cayley octonion or a split octonion (a.k.a., a bi-quaternion), taken to be specifically a split octonion. Thus, we now explicitly consider the bi-quaternion form and show how the formalism retains its desirable properties, explaining U(1)xSU(2)xSU(3) for example, while now becoming manifestly Lorentz invariant as well. In adopting the split Cayley algebra form of maximum information emanation zero-divisors occur that are removed to achieve maximal analytic continuation. Removal is an exact process on a complex product space (that is fully split to the chiral noise dimension of 29 'free' real dimensions). The effective dimension has been shown to be slightly greater than 29, however, here called 29^*. Knowing an effective dimensionality of a hypothesized maximal information emanation can be used in a number of arguments, including in conjunction with a hypothesis that such maximal information emanation be 'at the edge of chaos' – where the fractally optimum Mandelbrot set, with boundary dim=2 [38], is taken as a limit map description for each of the split dimensions that is complexified. The Feigenbaum universal constants [2] occur in the Mandelbrot set, the second of which is denoted here C_∞, which describes maximum perturbation 'at the edge of chaos'. A relation between $\{C_\infty, \alpha, 29^*\}$ is thereby indicated [37]. By considering the proposed maximal analytic continuation from 29^* real dimensions, we arrive at a derivation for an 'h-value' in the formalism that is very, very, small, but with a small dimensionful constant factor it becomes the familiar $\hbar$.

7.1 Introduction

A now obscure paper, published in 1917, describes Lorentz invariance for complex bi-quaternions [61] (modern notation [62,63]). In the description of Emanator theory thus far [1,6,15,16,22,37,46], a general octonion is indicated that can either be a regular Cayley octonion or a split octonion (aka, a bi-quaternion). In most of the analysis thus far the specific form of a Cayley Octonion is assumed, although the first paper [1] refers to "chiral bi-sedenions" in the title. The specific bi-sedenion formulation is what we return to here, because it leads to the manifestly Lorentz invariant form shown in [61]. This is because a split sedenion will be split at all sub-Cayley levels, thus selects the form of the chiral trigintaduonion with a split octonion [17]. The specific results of [1,6,15,16,22,37,46] 'carry over' with the shift from octonion to split octonion (the prior results typically involve term counting that is independent of sign, and all that changes in the multiplication rule when going from octonions to split-octonions is a single sign). What follows is a review of the above derivations, starting with a background section on the Lorentz transform, followed by background on the prior key results [1,6,15,16,22,37,46], with their (trivial) modification when shifting from Cayley to split-Cayley described. Notably, the product space argument leading to the explanation for $\hbar$ [46] was previously restricted to chiral emanators that were infinitesimal variations on unity (to allow decoupling at linear order). This restriction is no longer needed in the split formulation as *a full decoupling occurs by construction*. This strengthens the derivation for $\hbar$ significantly so it is placed in the Methods that follows and discussed further in the Results along with the Lorentz Invariant construction.

When examining the chiral trigintaduonion emanation at maximal dimensionality (that would preserve unit norm on the propagated trigintaduonion) we arrived at a template for the 10D chiral form. Recall that a general trigintaduonion T is 32D and can be written in terms of 4 octonions: T=((a,b),(c,d)), while the chiral T can be written in terms on 1 octonion and two real parameters: T=((a,α),(β,0)), where there are four chiral forms. The choice on the octonion 'a' in the chiral form was made in [1,6,15,16,22,37,46] to stay with the conventional Cayley algebra construction, not the split form. But either choice could be made, and the split form is arguably a larger space (thus a larger channel for maximum information emanation). The original hesitancy to use the split form is it has zero-dividers and other oddities not seen until operating in the full 32D trigintaduonion propagation, which seemed like something to be avoided .... but later analysis [3,7,46] revealed that removal of the ZD's provides a mechanism for introducing point-like matter. So there is no longer the aversion to having the split form. Furthermore, as indicated, if the ZD structure is not to be avoided, rather maintained, then when working with a chiral-T emanation within the larger T-space, where the larger space permits ZD's, it would be necessary to choose the split form in the T-chiral space to faithfully carry-over any ZD structure. Note that switching the type of octonion in the chiral form does not alter the independent octonion counting analysis underlying the results of [1,6,15,16,22,37,46], thus those results remain unchanged.

7.2 Lorentz transform Background
7.2.1 Lorentz transform notation
We can write the coordinate transformation under Lorentz transform as

$$x^{\mu'} = \Lambda^{\mu'}{}_{\upsilon} x^{\upsilon} ,$$

where $x^{\mu'}$ is a contravariant coordinate, and if the 4-vector length is invariant, then the metric is invariant:

$$\eta = \Lambda^T \eta \Lambda.$$

Any 4-vector entity z^{υ} that transforms as x^{υ}, is, similarly, called contravariant. This is to distinguish it from covariant 4-vectors that transform as:

$$z'_{\mu} = (\Lambda^{-1})^{\upsilon}{}_{\mu} z_{\upsilon}.$$

Note that for rotation transformations in 3D we have $R^{-1} = R^T$, so the '3-vectors' transform identically, and there is no differentiation of contravariant and covariant vectors. For the 4D Lorentz transform, however:

$$\eta = \Lambda^T \eta \Lambda \quad \rightarrow \quad \Lambda^{-1} = \eta^{-1} \Lambda^T \eta$$

and Λ^{-1} and Λ^T are not the same.

7.2.2 The 4-vector reformulation of EM – proof of Lorentz Invariance by construction [64]
If the 4-vector length is invariant under Lorentz transform, then so is the metric as indicated previously. Taken further, this invariance extends to any scalar or scalar product. The general transform properties for a full tensor generalization then follow. To show that the Maxwell equations describe a Lorentz invariant theory, we must identify 4-vectors that might be formed from the electric and magnetic fields. Such 4-vectors, if grouped (inner product contraction) to form a scalar entity, will be manifestly Lorentz invariant. If such scalar entities can be constructed that coincide with the Maxwell equations, then the proof of Lorentz invariance is complete by construction. Recall the component notation for the E and B fields in terms of potentials:

$$E_k = -\nabla_k \varphi - \frac{\partial A_k}{\partial t} \quad and \quad \epsilon_{ijk} B^k = \partial_i A_j - \partial_j A_i$$

So far we have the 3-potential A_k and need to generalize it to a '4-potential'. Furthermore, the 3-gradient operator has a natural generalization to a 4-gradient operator (that transforms like a covariant vector). If we identify the 4-potential to be:

$$A_\mu = \left(-\frac{\varphi}{c}, A_k\right),$$

then (setting c=1):

$$E_k = \partial_k A_0 - \partial_0 A_k .$$

The forms E_k and $\epsilon_{ijk} B^k$ suggest that we examine the antisymmetric 2nd rank tensor formed from the 4-potential by the relation:

$$F_{\alpha\beta} = \partial_\alpha A_\beta - \partial_\beta A_\alpha,$$

where we see the E and B components reside in this structure:

$$F_{\alpha\beta} = \begin{bmatrix} 0 & -E^1 & -E^2 & -E^3 \\ E^1 & 0 & B^3 & B^2 \\ E^2 & -B^3 & 0 & B^1 \\ E^3 & B^2 & -B^1 & 0 \end{bmatrix}$$

Let's consider a Lorentz transform corresponding to a boost in the x-direction (and no longer have c=1), then $\Lambda^0{}_0 = \gamma = \Lambda^1{}_1$, and $\Lambda^0{}_1 = -\gamma v/c = \Lambda^1{}_0$, and $\Lambda^2{}_2 = 1 = \Lambda^3{}_3$, with the rest zero. Let's see how the E an B components transform:

$$E'^1 = E^1 \quad and \quad B'^1 = B^1$$
$$E'^2 = \gamma(E^2 - vB^3) = \gamma(E^2 + (v \times B)^2)$$
$$etc.$$

These are indeed the transformations consistent with the coordinate transform of Maxwell's equations (can be shown by considering an infinitesimal Lorentz boost as indicated). Since we began with a potential formulation, this is then equivalent, without loss of generality, to a proof of Lorentz invariance for the homogenous Maxwell equations. For the inhomogeneous Maxwell equations we have

$$\rho = \epsilon\nabla \cdot E = \epsilon\partial_k F^{0k} \quad and \quad J^i = \epsilon\partial_k F^{ik} + \epsilon\partial_0 F^{i0}$$

thus we form the 4-vector for current to be:

$$J^\mu = (\rho, J^i),$$

which means that the inhomogeneous Maxwell equations can be written as a contravariant 4-vector expression (which is manifestly Lorentz invariant):

$$\epsilon\partial_\nu F^{\mu\nu} = \frac{1}{c} J^\mu.$$

Lastly, for the effect of an electromagnetic field on a moving particle we have the Lorentz equation:

$$\frac{dp}{dt} = q(E + v \times B), \quad p = \gamma m v.$$

To get to a 4-vector formulation we already know that p generalizes to the 4-vector:

$$p^\mu = (E/c, p).$$

On the RHS we have a $\boldsymbol{v} \times \boldsymbol{B}$ term. This suggests a tensor form involving $F^{\mu\nu}$ (to get the B^k term) and the velocity 4-vector $U^\mu = (\gamma c, \gamma v)$, for which we find:

$$q(\boldsymbol{E} + \boldsymbol{v} \times \boldsymbol{B}) \to \frac{q}{\gamma c} F_{j\mu} U^\mu$$

The 4-vector form of the Lorentz equation then becomes:

$$\frac{dv}{dt} = \frac{q}{\gamma c} F_{\nu\mu} U^\mu.$$

which is manifestly Lorentz invariant. Thus the Maxwell equations and the coupling of the EM Fields to charged matter are all Lorentz invariant. (Note that charge is a scalar, thus is seen to be the same in all Lorentz frames.)

7.3 Appearance of Lorentz-invariant Spinor Solutions in addition to Vector solutions [65]

A mathematical subtlety that arises at this juncture is not due to the novel three boosts introduced by the 4D Lorentz transformation, but due to the three (standard) rotations. This is because the rotation group in 3D, SO(3), admits a double cover from the group SU(2), indicating two types of solutions. We find that there is a new type of Lorentz invariant description, other than the 4-vectors, known as a spinor. The easiest way to show that Lorentz invariance extends to the new mathematical object is to rewrite a 4-vector as a 2x2 Hermitian matrix, where V^a with components (V^0, V^1, V^2, V^3) is written as the 1-1 mapping as

$$\psi(V^a) = V^{AA\prime} = \begin{pmatrix} V^{00\prime} & V^{01\prime} \\ V^{10\prime} & V^{11\prime} \end{pmatrix} = \frac{1}{\sqrt{2}} \begin{pmatrix} V^0 + V^3 & V^1 + iV^2 \\ V^1 - iV^2 & V^0 - V^3 \end{pmatrix}$$

where the length of the 4-vector is related to the determinant of the matrix:

$$\det[\psi(V^a)] = \frac{1}{2} \eta_{ab} V^a V^b.$$

Thus, there is a map from SL(2,C) to the Lorentz transforms where there is a 2-1 isomorphism. This can be seen to be directly related to the SU(2) double cover (2-1 mapping) on SO(3) as mentioned previously (we consider the Lorentz transform leaving the time-like component of a chosen orthonormal tetrad invariant, what remains is the 3D rotation group SO(3)). This is not simply a useful mathematical relation. We've identified the invariance under Lorentz transform as a fundamental element of the theory (giving rise to special relativity, etc.). *If we see this then conveyed to an extended set of spinorial solutions, in addition to vectorial, then new forms of energy/matter are indicated for the spinorial* *solutions. This is precisely what is observed: matter is spinorial (spin $\pm 1/2$) and (force) fields are vectorial.*

7.4 Lorentz invariance via complex biquaternions [61-63]

Let's now consider a similar process involving transformational invariance but instead of encoding the Lorentz transform in the form of matrix transformation invariance let's use elements of the Cayley algebras instead. Specifically, consider the following transformation:
$$q' = aqa_c^*, \ where \ \ aa_c = 1,$$
where a is a (unitary) complex bi-quaternion: $H(\mathbb{C}) \times H(\mathbb{C})$, and $q = (ct, ix, iy, iz)$ (note this notation is for $q = (ReH_1, ilmH_1, iReH_2, ilmH_2)$). The $q' = (ct', ix', iy', iz')$ that results will correspond to a proper orthochronous Lorentz transform [61-63].

This is a remarkable result, but is there better (higher dimensionality/complexity)? Consider that a complex bi-quaternion is isomorphic to a complex octonion which is isomorphic to a 'chiral' sedenion, thus a theory for chiral sedenion emanation is indicated from this result as far back as 1917. The 'halting condition' on the generalization in 1917 seems to be that octonions are the highest-order of the division algebras that have inverses defined (which is necessary to have $aa_c = 1$ be defined). But they have already extended past octonions since these are complex octonions $\cong$ chiral sedenions, so how are they guaranteed to have $aa_c = 1$ be defined? This is possible for the *chiral* sedenions, as shown in [1], if restricted to be unit norm. So now we have our answer based on the results from [1] – we can go one complexation order higher:
$$Q' = AQA_c^*, \ where \ \ AA_c = 1,$$
where A is a (unitary) bi-complex bi-quaternion: $H(\mathbb{C} \times \mathbb{C}) \times H(\mathbb{C} \times \mathbb{C})$, which is isomorphic to a unitary quaternionic bi-quaternion, $H(\mathbb{H}) \times H(\mathbb{H})$, and $Q = (ct, ix, iy, iz)$.
where $Q = (ReH_1, ilmH_1, iReH_2, ilmH_2)$ as before, except $H_1 = H \times \mathbb{H}$ not $H_1 = H \times \mathbb{C}$
The $Q' = (ct', ix', iy', iz')$ that results will again correspond to a proper orthochronous Lorentz transform. Again, how do we know the critical operation $AA_c = 1$ can always be satisfied? Previously we saw that a complex bi-quaternion was equivalent to a 'chiral' sedenion (in the sense described in [1]), thus a bi-complex bi-quaternion is isomorphic to a complex chiral sedenion, which is isomorphic to a 'doubly chiral' trigintaduonion. This is precisely the construct examined in [1], so a generalization of the 1917 result to unitary quaternionic bi-quaternions

appears possible. As shown in [1], however, there is no higher order construct. This latter form (on doubly-chiral trigintaduonions) establishes the Emanator as Lorentz Invariant explicitly. To consider the 1917 relation (involving chiral sedenions) directly in the analysis, we shall simply expand the doubly chiral trigintaduonion to a (singly) chiral sedenion and operate at that level in the analysis.

7.5 Lorentz invariance via complex biquaternions using Split Octonion Algebra

Maximum information flow occurs for octonions when split, because additional structure can occur, such as zero divisors. Maximum information flow with a propagate-able mathematical structure is thus a chiral trigintaduonion of the form

$$T_{chiral}^{(k)} = \begin{cases} ((O_{split}, \alpha), \beta) \\ ((\alpha, O_{split}), \beta) \\ (\beta, (O_{split}, \alpha)) \\ (\beta, (\alpha, O_{split})) \end{cases}.$$

Point-like matter appears in the theory when the zero-divisors in the propagation are managed by removal according to maximal analytic domain (the maximum flow hypothesis will occur in a maximal domain). Thus, it is found that the zero-divisor general formulation of the maximal propagator is indicated by maximum information flow and indicates point-like matter. This formulation involves the split-Cayley algebras. Thus, the chiral trigintaduonion is a chiral bi-sedenion as originally indicated [1]. The chirality at sedenion-level is:

$$S_{split,L} = (O_{split}, O_{real}) \text{ or } S_{split,R} = (O_{real}, O_{split}).$$

If we are working with a chain of products of chiral emanator trigintaduonions $T_{chiral}^{(1)}$ this will reduce to a chain of products of $((O_{split}, \alpha), \beta)$ and upon using the unit-norm constraint to eliminate β as an independent variable, we have a product of (O_{split}', α'). So, dropping the primes, we have products:

$$(O_{split}, \alpha)_1 \times (O_{split}, \alpha)_2 \times \dots$$

Since $O_{split} \cong H \times H$ and $(0, \alpha) \cong O(\mathbb{C})$, we have

$$(H(\mathbb{C}), H(\mathbb{C}))_1 \times (H(\mathbb{C}), H(\mathbb{C}))_2 \times \dots$$

Recall in the background section that if we wanted to consider the standard equivalence class of invariants for unitary complex bi-quaternion operations, then we call

$$(H(\mathbb{C}), H(\mathbb{C}))_1 = a_c^*, \quad aa_c = 1, \quad |a| = 1,$$

and have propagation according to:

$$qa_c^* \to q_{new}$$

97

and invariance according to
$$q' = aqa_c^*$$
where invariance for $q = (ct, ix, iy, iz)$ is realized such that the $q' = (ct', ix', iy', iz')$ that results will correspond to a proper orthochronous Lorentz transform. The maximal information hypothesis underlying the proposed Emanator theory, thus indicates Lorentz invariant point-like matter.

Chapter 8. Manifold Thermality from a dual complex 'contact' structure, and Manifold as Apparatus

8.1 Universal Thermality

There is a fundamental complex structure that still remains from the Emanator formalism upon projection from the higher-dimensional Cayley algebras into the maximum propagation described, for the alpha maximum-perturbation trigintaduonion. That complex structure is 'contact' analytic and is realized under conditions where limits involving that structure are taken to zero (or to some fixed value at component-level). Two main examples where this has been most significant are in (i) shifting a QFT to a thermal QFT by shifting to imaginary time related to the inverse temperature [66]; and (ii) in the use of dimensional regularization to renormalize QED and QCD [67]. Thus, the process of emanator theory settling on the maximum information propagation dimension is one where higher order complex structure (from non-propagate-able dimensions) is still accessible for regularization processes. A possible *origin* for the complex structures is described in the next section.

The significance of the remnant complex structure is ***universal thermality***. Thermality via complex structure and Lorentz Invariance on chiral Trigintaduonion are more fundamental constructs than QFT (based on the Standard Model) or GR (with manifold dynamics according to the Standard Cosmological Model). This is because the limits defining such thermality would exist for a single chiral Trigintaduonion propagation step, and we don't 'see' QFT until enough emanation steps have occurred such that the product group gauge that begins to resolve is the gauge group of the standard model, thus giving rise to the QFT of the Standard Model.

Universal thermality explains the odd behavior seen upon 'complexification', such as BH geometry complexification giving the Hawking temperature of a Black Hole [68] consistent with the full quantum field theory analysis at the BH horizon demonstrating thermal emission with that same temperature done by Hawking [69]. Given the Hawking radiation, a BH is clearly unstable and will evaporate over time.

If the BH is 'placed in a box' or has not-flat asymptotic geometry, however, then the 'closed' system may reach an equilibrium and be stable. This is precisely what is considered by Hawking and Page (1983) [70] upon examining the thermodynamics of AdS BH's. Further work along these lines was done in [71-73]. In these efforts a Hamiltonian formulation for Gravity is made and then complexification yields a partition function describing an ensemble of BH solutions. Stability, in the form of positive heat capacity, can then be shown.

8.2 Doubly Chiral Split Octonion Emanation with a Doubly Complex Structure

The doubling limits on extension, as well as the extensions themselves, are artefacts of the Cayley algebra construction that is employed. The original double-chiral extension was done to extend the unit norm propagation property from octonions to the trigintaduonions [1]. Here is the emanation expression in various forms:

$$T_{base,new} = T_{Em}\{T_{base,old}\} = \sum T_{base,old} T_{chiral}^{(k)} = T_{base,old} \cdot T_{Em,Eff}.$$

In the last form we speak of an effective emanation ($T_{Em,Eff}$) off of the original base trigintaduonion. In the explicit sum form we see that emanation is a sum over chiral multiplications, where a chiral trigintaduonion has the distinctive 'doubling' from of extension:

$$T_{chiral}^{(k)} = \begin{cases} ((O_{split}, \alpha), \beta) \\ ((\alpha, O_{split}), \beta) \\ (\beta, (O_{split}, \alpha)) \\ (\beta, (\alpha, O_{split})) \end{cases}$$

Consider the first chiral emanator form: $((O_{split}, \alpha), \beta)$. In [1] we began with unit-norm propagation on octonions:

$norm(O_{base}O_{propagator}) = 1$ if $norm(O_{base}) = 1$ and $norm(O_{propagator}) = 1$

and extended via the Cayley hierarchy two level higher to the trigintaduonions:

$norm(T_{base} \cdot T_{Em,Eff}) = 1$ if $norm(T_{base}) = 1$ and $norm(T_{Em,Eff}) = 1$.

By going two-level higher we can still expand the Cayley products two-levels down, back to the level of tri-octonionic products, where the norm operation is defined and associative in the sense that $norm(O_1 O_2 O_3) = norm(O_1)norm(O_2)norm(O_3)$ (true for division algebras such as the octonions).

100

What is proposed now is a similar extension to that of $O \to ((O, \alpha), \beta)$, where the norm property (and associativity under norm) was extended. Here we want to extend:

$$T_{Em,Eff} \to \left(\left(T_{Em,Eff}, \gamma\right), \delta\right).$$

Before we were trying to extend a norm property with implicit norm associativity on that operation, here we merely want to extend the theory in a well-defined way that captures "maximal information emanation". At this juncture it would be nice to assume or impose a selection process on the product of effective emanations that is explicitly associative:

$$T_{base,old} \cdot T_{Em,Eff1} \cdot T_{Em,Eff2} \cdot T_{Em,Eff3}$$
$$= T_{base,old} \cdot T_{Em,Eff2} \cdot T_{Em,Eff3} \cdot T_{Em,Eff1}$$

but to do so would clearly be limiting on the "maximal information flow" by dropping the non-associativity (think of it as dropping 1D of information). In the context given, however, we are extending via two layers of complex structure indicated by the parameters γ and δ *(so extending by 2D for a net gain of 1D)*. Thus, we hypothesize that maximal information flow will select effective chiral emanations to be associative such that two layers of complex structure can be carried by the maximum perturbation emanation (where $\gamma \to 0$ and $\delta \to 0$ to then give the 'contact' complex structure indicated above. The selection for associativity may dictate the 'size of the hand dealt' in the chiral emanator deck discussion, where the number of cards, e.g. a sum of chiral emanation steps (78 or 72), are 'dealt' until associativity vis-vis the preceding two achiral (summed) emanations steps is achieved. Further elaboration on these ideas will be made in a later paper.

8.3 Apparatus, Measurement Theory, and related Emanation Processes

In work involving a full general relativistic solution to describe dust shell collapse it was found that the geometry is effectively "pure apparatus" [74]. Geometry is thus at least part of the apparatus. Consideration of 'bare' QFT in an arbitrary apparatus leads to point like violation of the weak energy condition (WEC), but with choice of apparatus can have average WEC (AWEC) such that no violation of AWEC is seen. The separation of quantum observable and classical apparatus has always been a troublesome complication; but in Emanator theory this is simply seen as a fundamental aspect of the emanation process. Consider repeated 'emanation' steps starting from Tbase, we then have two aspects: Tbase vs the product algebra (Tbase x Tem) x Tem x … Consider these objects:

Tbase: this is a higher-order spinor, spinor fields describe (quantum) matter
(Tbase x Tem) x Tem x ... : gives the standard model gauge field that is Lorentz Invariant

From the Lorentz Invariance on 4D we get local Minkowski space-time, which then generalizes to a manifold globally. So, geometry appears in this process in the context of the product algebra (taken to be equivalent to apparatus) and the matter appears in spinor form (taken to be a quantum observable).

Chapter 9. The Fractal-G Hypothesis

9.1 Introduction

Strong experimental evidence is established for the achiral emanation process to be Martingale. The significance of this is that statistical mechanics processes that are Martingale can have equilibria and other well-defined limit phenomena, precisely what is needed for emanator theory to describe a physical theory. Well-defined limit processes, together with the maximum noise propagation dimension of 29* derived in [22,76], indicates that the oddities of the Dirac Large Number Hypothesis [75] should be re-examined. In doing so we arrive at the fractal G hypothesis, which explains the origin of the gravitational constant.

Computational results on achiral emanator processes with over 10^9 computational emanation steps was initially reported in [37]. Further evaluation of those results show that the achiral Trigintaduonion evolution, component-wise, is exactly like a random walk. A random walk process is Martingale [77], thus strong experimental evidence is established for the achiral emanation process to be Martingale. The significance of this is that statistical mechanics processes that are Martingale can have equilibria and other well-defined limit phenomena, precisely what is needed for emanator theory to describe a physical theory.

The Dirac Large Number Hypothesis [77] has been a mystery that has been noted by many other physicists [78-88]. At issue is the odd occurrence of a lot of ratios of different 'fundamental' lengths in the theory, that tend to group in families where their ratio is 10^{40}, or 10^{60}, or 10^{80}, etc. (depending on how you want to group the terms). This 'fractal' behavior at large scale (a classic fractal trait being repeating structures at different scales) has not been couched as a fractal phenomenon previously because there wasn't a context to warrant such a supposition. But here we know the maximal information emanation hypothesis is likely to force the emanation process to operate at the edge of chaos, where the process becomes fractal or has a well-defined fractal limit (given the Martingale property), such that we arrive at the fractal G hypothesis, which will explain the origin of the gravitational constant.

9.2 Existence of generalized unit-norm propagation structure that is 10D

Recall from previous section that unit-norm right product propagation is trivial for the division algebras since norm(XY) = norm(X) × norm(Y). From this it is apparent that we have an automorphism group given by the norm itself (A(XY)=A(X)A(Y)), and in the case of the octonions this automorphism group is G2 [20]. It can be shown that SU(3) is in G2 [20]. Let's now consider the situation with a higher-order Cayley algebra, the Sedenions, 'S'. We obviously don't have norm(S_1S_2) = norm(S_1) × norm(S_2) in general, as this would then allow S to join the ranks of the division algebras, and it is proven that such don't exist above the Octonions [21]. Can we still have a propagation structure? Is it possible to have a 'base' sedenion for which norm(S_{base})=1, and to have a right propagator (product) sedenion also norm(S_{right})=1, such that norm(S_{base} x S_{right}) =1? The answer is yes (see appendix of [22] and [1]), when the sedenion has the (chiral) form of an octonion crossed with a real octonion: S_{chiral} = (O,O_{real}) or S_{chiral} = (O_{real},O). Can we continue this to arrive at a propagation structure on the Trigintaduonions? Again the answer is yes, with the chiral form generalizing off the chiral Sedenion as might be expected: T_{chiral} = (S_{chiral}, S_{real}) or (S_{real}, S_{chiral}) [1]. It is proven that this extension process will go no further [1]. What happens is that due to the chiral form we are still able to re-express all T products (or S) as collections of terms involving tri-octonionic products (which have nice properties as described in [1]), and this can no longer occur above the (chiral) trigintaduonion level.

Note that most other analyses of propagator constructs stay within the division algebras, thus never go past the octonions, or past 8D. In the excellent documentary "Spinal Tap" they describe their amps going to '11', not just 10 like everyone else. Mindful of that, my propagator goes to 10D, not just 8D or less. So some interesting things are bound to happen.

9.3 The Dirac Large Numbers Hypothesis
Recall the hypothesized relation [37]:
$$\alpha^{-1} = \left(\sqrt{C_\infty}\right)^{29^*}, \quad where \ dim. = 29^*.$$
Here we are imagining maximum information flowing via 29* effective dimensions, where the phenomenology is concerned with the *number* of

dimensions and arrives at a dimensionless parameter α. We might as readily ask for the maximum information that would flow through the 137* independent tri-octonionic terms at chiral-trigintaduonion component level. Here the phenomenology is concerned with the amount of 'noise' that each terms could carry to arrive at a dimensionful parameter 'X':

$$X^{-1} = \left(\sqrt{C_\infty}\right)^{137^*}$$

Evaluation leads to

$$X = 8.05077 \times 10^{-11} [dimensionful \ unit].$$

In SI units the gravitational constant G is:
$$G = 6.674 \times 10^{-11} [N \ m^2 \ Kg^{-2}],$$
so, is the nearness of these numbers a coincidence? Taken alone, yes certainly, but taken together with a collection of other odd 'coincidences', noted over the past 100 years [78-88], maybe there's something to this.

Consider relation that motivated Dirac to suggest the large number hypothesis. Working with 'constants' that are seen in the micro and macro scale, such as the Hubble constant H, with age of the universe $\sim H^{-1}$, the mass of the nucleon (proton) m_p, we have the following:

$H^{-1} \cong 4.3 \times 10^{17}$: *the age of the universe* (macro-scale constant)

$\tau = \dfrac{e^2}{(m_e c^3)} \cong 10^{-23}$: *the strong time scale* (micro-scale constant)

where m_e is the electron mass, c is the speed of light, e^2 is the electrostatic coupling between a single electron and proton, for example, e.g., exhibiting electrostatic force e^2/r^2. The gravitational force between an electron an proton, on the other hand, is $Gm_e m_p/r^2$:

$\dfrac{Gm_e m_p}{r^2}$: *gravitational force* (macro-scale)

$\dfrac{e^2}{r^2}$: *electrostatic force* (micro scale)

Consider another ratio of 'fundamental' constants operational on some system that spans the micro and macro realms (e.g., the Universe vs internal, or the macro gravitational system versus its micro system of hydrogen). Consider the mass of the observable Universe M in a ratio with the nucleon mass, which gives $\sim 10^{80}$, thus the square root is used to group it with the $\sim 10^{40}$ ratios. There are going to be obviously going to be large number ratios if we take "universe" / "elemental particle" in some operation, so this isn't what is surprising. What is odd is that these large numbers ratios appear to group as $\sim 10^{40}$, $\sim 10^{60}$, and $\sim 10^{80}$. For the above cases we have:

$$\sim 10^{40} \approx \frac{H^{-1}}{\tau} \approx \frac{e^2}{G m_e m_p} \approx \sqrt{\frac{M}{m_p}}\,.$$

9.4 Fractal G Hypothesis

If we take the Dirac relations as part of the fractal nature of reality (with optimization pushing to "the edge of chaos" and inducing fractal effects), with repeating structure at different sales and similarity relations between 'things' at different scales a possibility. Given this perspective, interpreting the constant 'G' as a multiscale fit parameter across all of these domains, such that the single observed G suffices, is an interesting prospect. Note that not all the terms in the ratios involve G so we can solve for G, approximately, accordingly.

Thus, the parameter we know as G is hypothesized to actually be a fractal fit parameter, albeit still one of the 22 parameters determined in the emanation process. G is not like the other 22 constants in that it doesn't describe elementary particle mass, say, but gives a parameterization of the fractal structures that occur at different scales when coupling matter to geometry. In the derivation of $|h^*|$ we see that maximal information flow occurs 'at the edge of chaos', i.e., the description is fractal, so this is consistent. As with the $|h^*|$ derivation capturing the extreme smallness of Planck's constant h up to the actual constant chosen to give precisely h ($h = c|h^*|$), here we have the significant smallness of the gravitational constant captured in the term X up to the actual constant chosen to give precisely G (e.g., $G = kX$).

Chapter 10. The Emanation is Martingale Hypothesis

In this chapter we show computational results (Sec. 10.1) that indicates that the (achiral) Emanation process, at component-level, behaves as a random walk (asymptotically). In Sec. 10.2 we discuss the implications of this for the Emanation process, since true random walk processes are provably Martingale.

10.1 Random Emanation-Walk Results

$T^{(1)}$chiral emanation: $((0, \alpha), \beta)$ form, with noise δ at the indicated template positions aside from T[0] component, which is ~1 with unit-norm normalization (where all other components, if non-zero, involve a max $\delta/2$ noise, noise uniformly distributed $\pm|\delta/2|$). Emanation is then simply multiplication: $\left(\mathbf{T} \bullet T^{(1)}_{chiral} \right) \bullet T^{(1)}_{chiral} \bullet \bullet \bullet$, where here we see how many emanation steps it take to go from T[0]=1 in the initial base trigintaduonion to T[0]=0 (the number of steps to the first zero-crossing). These are effectively random walk simulations on the unit-norm trigintaduonion subspace S^{15}, where the emanation step is chiral (see Table 1).

δ	N_{avg} (5 samples)	$\left(\frac{\delta}{2}\right)\sqrt{N_{avg}}\sqrt{2/\pi}$
0.7	9.0	0.8376
0.6	17.2	0.9932
0.5	32.4	1.1368
0.4	34.2	0.9334
0.3	75.2	1.0371
0.2	147.8	0.9693
0.1	706.0	1.061
0.05	2819.4	1.057
0.01	43,136.0	0.8297
0.005	206,454.4	0.9055
0.002	1,613,224.8	1.0131
0.0016	3,532,666.8	1.1997

Table 1. $T^{(1)}$ chiral emanation random-walk simulation.

Let's now consider off-template $T^{(1)}$chiral emanation: $((0, \alpha), \beta)$form with T[7] and T[24] swapped (thus have δ noise off template, which breaks the unit-norm preserving property in the emanation multiplication $\left(T \bullet T_{chiral}^{(1)*}\right) \bullet T_{chiral}^{(1)*} \bullet \bullet \bullet$, where the * denotes the off-template form (see Table 2).

δ	N_{avg} (5 samples)	$\left(\frac{\delta}{2}\right)\sqrt{N_{avg}}\sqrt{2/\pi}$	norm(T) at zero-crossing
0.7	12.0	0.9674	~1
0.6	14.6	0.9146	~1
0.5	19.4	0.8786	~1
0.4	19.0	0.6956	~1
0.3	66.4	0.9752	~1
0.2	254.2	1.2721	~1
0.1	721.8	1.0718	1.0179
0.05	3103.2	1.1112	1.0011
0.01	46883.6	0.8638	1.0001
0.005	213,397.0	0.9214	1.0009
0.002	975,838.4	0.7882	1.0001
0.0016	2,248,293.0	0.9571	1.0001

Table 2. $T^{(1)*}$ chiral emanation random-walk simulation.

Let's now consider an Emanator definition that involves a 4-suit (chirality) generation process that is summed and renormalized to 1 at each step (to be achiral). The 'deck' of four cards (general chiral class members) that is summed leads to a modification of the ranwalk equation:

$$ranwalk(\delta, N_{avg}, |deck|) = \delta\sqrt{N_{avg}}\frac{\sqrt{2/\pi}}{\sqrt{|deck|}} = \delta\sqrt{\frac{2N_{avg}}{\pi|deck|}}$$

The results for 4-suit Emanation are shown in Table 3. If the perturbation is generated in a range (uniformly in$[-\delta..\delta]$) it has half the step-size (on average) and has possible mixing of chiral emanations that are within the perturbative limit and not within the limit.

δ	4-suit with[-1..1] N_{avg} (5 samples)	*ranwalk* ($\frac{\delta}{2}$ on avg.)	4-suit with[-1,1] N_{avg} (5 samples)	*ranwalk*
0.1	2583.0	1.013775	772.0	1.108457
0.05	11584.0	1.073445	2926.6	1.079100
0.01	246437.8	0.990220	144019	1.513979
0.005	792124.4	0.887660	347164.2	1.175297
0.001	15049973.4	0.773834	6079663.0	0.983671
0.0005	-----	-----	22,486,524 (one)	0.945891
0.00025	-----	-----		
0.000125	-----	-----	>100M	

Table 3. 4-suit Emanation Steps. Random walk transitions from noise additivity in quadrature to linear. At transition to analyticity have $\sqrt{2}$.

In Table 3, the [-1,1] case shows possible transition at 0.01: into "analytic" domain, where random walks explore more terrain, have more movement, get possible $\sqrt{2}$ effect, then transition to linear perturbative domain. A transition from quadratic to linear noise dependency is becoming apparent.

In Table 4 is shown the results for when the Emanator Deck is 72, with 4 sums to get the different $\pm\alpha$ and $\pm\beta$ chiral templates, and a different four sums associated with the 4 'suits' or chiralities. Consider linear noise additivity within the noise components of a given chirality during the chiral emanation:

$$ranrun(\delta, N_{avg}, deck) \propto \delta N_{avg}$$

Shown in table results from runs with 72-deck with noise drawn from δ x [-1,1]. A clear linear relationship exists. The same strong linear relations exists for δ x [-1..1] based emanation, but proceeds more slowly, so dataruns not as complete and not shown. The result "with Major 1" has noise injection δ at the position of the 1st Major perturbation (as in major arcana, since similar card subgroups as in the construct of the tarot deck). The result with "Tarot" Emanation uses the 72-deck with random noise injection according to the probability of a "card" from a 78-style emanation "card" tarot deck (this appears to be the most complete case for achiral emanation with the full range of non-chiral perturbations allowed).

δ	72-deck [-1,1] N	72-deck [-1,1] N With Major 1	Effective 78-deck "Tarot" Emanation
0.1	212	214	422
0.05	427	428	851
0.01	2137	2133	4298
0.005	4274	4264	8531
0.001	21372	21353	42740
0.0005	42745	42751	85476
0.00025	85493	85572	170961
0.000125	170986	171139	---------

Table 4. 72-deck Emanation Steps. Random walk in linear noise additivity regime.

The repeated experiments show remarkably small difference in the 72 deck counts even with non-pathology cases [-1..1] and outside mixing domain (if we are even seeing one) – i.e, all 72 deck runs appear to be in the perturbative regime, with linear growth seen for the entirety of the Results in Table 4.

The computational results shown above confirms that noise, or random walk steps, add in quadrature, thus $\propto \sqrt{N}$ distance, until an analytic perturbation regime reached, where noise then adds linearly, thus random walk goes $\propto N$ distance with N steps.

10.2 Emanation is Martingale
In analysis of zero-crossing events in Tbase (a necessary condition for a zero-divisor event with a Tchiral product) [37] it was noted that the achiral behavior precisely matched, component-wise, that of a random walk towards the zero-crossing event. Experimentally, this shows that the Emanation process is a Random Walk process. But Random Walk processes are known to be Martingale [77], thus Emanation is Martingale. Systems that are Martingale have limits, such as the familiar equilibrium limit. The hypothesis of universal thermality (see Disc.) is consistent with this existence of equilibria result.

Emanator theory projects to a quantum theory with a complex propagator and associated Action functional theory. This is proposed to be a reified (dimensionful units) emergence that is guided by the dictates of the quantum deFinetti constraint (to have a complex propagator [19]);

analyticity (for maximal continuation with maximal information flow); and the product-algebra gauge group of the standard model with representation and particle families according to the extended standard model that includes massive sterile neutrinos (to arrive at 22 parameters). The emanator projection is seen as a fundamental aspect of the process. In this context, the suggestion of thermal universality emergence with analytic time may not seem so extreme.

Chapter 11. The Emanation Process

11.1 The process

Consider numerogenesis from an infinite-order hypercomplex unit-norm 'Number' and 'Emanation' process (algebraic multiplication) giving rise to a propagating structure, with time and chirality self-selected, with QED and QCD gauge bundles emergent, for example, with their associated parameters fixed (including α). With the proposed chiral trigintaduonion emergence have 10dim propagation with α-perturbation into the full 32 dimensions. Thus, have a hypercomplex Big Bang with emergence of unit-norm base elements and the unit-norm resulting emanation step (or path sum). The receiving of the universal emanation results in emergent spacetime and chirality, perhaps akin to the emergence of Amman bars and orientation with a Penrose tiling once seeded [89].

The emanation construct is self-selected to have maximum information flow beginning with selection of the maximum dimensionality algebraic subspace for states (Step I below singles out chiral trigintaduonions and with perturbation limit $\{\alpha\}$ that is fractal.). Next is allowing the 10D chiral subspace emanation element to have maximum perturbation into the surrounding 32D trigintaduonion algebra. As mentioned in the Kato-Rellich and noise-budget analysis in the Methods, and related Results, maximum noise transmission is from unit-norm trigintaduonion base with right multiplication by emanation involving a unit-normal, maximal perturbation, chiral trigintaduonion. This results in the $\{\alpha, \pi\}$ relation (Step II), where maximum perturbation occurs when noise has maximum antiphase (thus introduction of 'π'). Next is allowing for emanation processes that are achiral, but composed of the high dimensional flow chiral elements, that are then summed (Step III). In other words, a fundamental sum on emanation paths is posited in the emanation process even if the paths are only a single step long (possible path conventions are discussed later). Here are the three steps:

> **Step I:** Selection, or projective emergence, of maximum-dimensionality subspace emanation process (10D) operating within its Cayley algebra (32D). By allowing the maximum allowed perturbation into the surrounding 32D algebra of elements, a fractal limit can be probed, as done with the

Mandelbrot set images, this limit reveals $\{\alpha\}$ purely computationally (theoretically this is related to the universal c_∞ derivation, as will be quantified at Step III).

Step II: Selection, or emergence, of emanation from Step I with maximum allowed perturbation into the surrounding 32D algebra of elements, where analyticity is assumed (it is part of the optimal selection process in the emergence from the higher dimensional hypercomplex numbers). The property of analyticity allows application of the Kato-Rellich theorem in related domains, and lays the foundation for Euclideanization and dimensional regularization (QFT renormalization) methods later. At this step, the noise-budget analysis is only based on the structure of the chiral trigintaduonion elements T and the structure of their right multiplication on a norm=1 base trigintaduonion (that is being emanated to a new base trigintaduonion), e.g. the structure T•T, from which the $\{\alpha, \pi\}$ relation is obtained. At this stage in the Emanator construction described in the Methods/Results we see a fundamental hypothesis of maximal noise when phase angle and imaginary component magnitude are equal (a Euclideanization, or analyticity, type relation).

Step III: Selection, or emergence, of emanation from Step II with maximum allowed perturbation into the surrounding 32D algebra of elements, at the boundary where analyticity is not assumed, but where an iterative mapping is induced with resulting universal limit properties from [2], giving rise to an effective dimension analysis for the iterative mapping, and thus a new relation "from the edge of chaos": $\{\alpha, \pi, c_\infty\}$. This analysis can be developed even further, since Kato-Rellich is used in the Methods to argue that there are no zero-divisors for perturbations up to the α limit. If we attempt to probe a little further, we start to encounter zero-divisors. Consider the limit density on zero-divisors at the α limit (taken from greater than α) how might this relate to quantum properties and the distinctive quantum constant h?

The emergence process with analyticity helps explain the validity of the various renormalization methods (dimensional regularization, in particular). In the latter regard, the dimensional regularization trick whereby a higher complex dimensional extension is invoked is here seen to actually be true. Similarly, string theory is an emergent construct, along

114

with the manifold and the standard model, and Lagrangian encapsulations, etc. Thus, invoking a higher dimensional space, often through complexification of real variables, is natural in this emergent from a higher hypercomplex algebraic space context, since a higher dimensional complex embedding is already posited to exist in the emanation emergence process. The complex-extension method is critical in QED, Euclideanized path integral formulations, and thermal quantum field theory in general, where complex time relates to introducing a thermal background temperature for the system (thus the complex extension allows unification with thermal physics and emergent, Law of Large Numbers based, statistical mechanics constructs).

In [9], with split octonions alone it is possible to describe spacetime, EM-fields, and uncertainty relations… This is very promising as regards extracting the familiar standard model from the much larger, already chiral, 10D propagation with maximal perturbation α (and 22 parameters from the non-propagating dimensionalities [15]). From this we get complete propagation with 78 generators (consistent with string theory, as is the 10dim). Also, we shall see that we have 137 tri-octonionic 'braids' of information flowing in the 10dim chiral propagation, this was critical in the derivation of π from α.

Just from the propagation structure on one path we have already seen core emergent structure that results in a universal emanation with structural parameters 10,22,78,137 and perturbation maximum $\alpha=\sim 1/137$. The central notion in the universal emanation hypothesis is that there should be maximal information flow, where this is accomplished by finding the highest theoretical dimensionality of unit-norm 'propagation', here called an emanation, which turns out to be 10, then add the maximal perturbation that still allows unit-norm propagation, where that perturbation is into the space the 10D motion is embedded in, here a 32 dimensional (trigintaduonion algebra) space.

Given maximum information flow, the universal emergence will arrive at the 10D propagation splitting (compaction) into spacetime geometry and matter gauge fields. The parameters and structure described are consistent with quantum field theory, where we fundamentally arrive at emergence of 'propagation' as conventionally known, with a complex Hilbert Space. A complex Hilbert Space description is the only one with propagation (details below), thus it is necessarily the emergent construct that must encapsulate the geometry/matter split/compaction, into the familiar

115

Standard Model formulations. This ties into emergence of the standard formalisms of QED and QCD. Likewise for the emergence of elegant geometrically optimal solutions relating to General Relativity (GR). Where there was conflict between QED/QCD and GR, e.g. the question of Quantum Gravity (QG), it will be solved by considering the universal emanation of not just one path but all paths, summed with the usual phase cancelations down to a 'classical path' with stationary phase. The latter, in this context, is the emergence of standard propagator theory with standard model. So proposing here an earlier phase of universal evolution described by a theory of emanations, where mathematically invariant emergent structures appear. From this early phase, one of the emergent constructs is the familiar path integral based on standard (unitary) propagators in a complex Hilbert space.

The implication of an emergent phase of universal evolution with standard propagators, etc., is not only a framework within which to possibly answer the questions of quantum gravity, but also a framework where the emergent trajectory has emergent 'time' (and parameter h, and euclideanization/thermality). In the end, the Black Hole (BH) conundrum in quantum gravity might reduce to a scattering calculation once 'boundary terms' are understood. With reference to the originating 'emanator' construct, we have a higher level second quantization but not based on standard propagators, but emanators. This new type of second quantization might shift to a notation where the trigintaduonion (bi-sedenion) structure dominates.

To recap: α, 10,22,78,137, are parameters resulting from analysis on a single path construct, where the number 22 corresponds to the number of emergent parameters in the description of the propagating construct. In addition, the time choice is emergent via a multi-path construct, along with the propagator construct, and is coupled in both time step (by h) and imaginary time increment (with Euclideanization regularization 'built in'). The formulation is inherently embedded in a higher dimensional complex space, thus all of the QFT complex analysis analyticity methods are valid as the assumptions made are now part of the maximal information flow emergent construct.

11.2 Maximal Information Propagation requires a complex Hilbert Space [19]

As mentioned previously, according to [19], a complex Hilbert space is selected by the quantum deFinetti theorem, since it is required for

116

information propagation (and thereby consistent with the maximum information propagation concept in its selection). Because it's a complex Hilbert space, this explains why the path integral operates in a complex space, even though the underlying universal algebraic construct from which it is emergent is hypercomplex to the level of the trigintaduonions.

From [19], a simple derivation shows why the quantum deFinetti Theorem requires amplitudes to be complex. Suppose $f(n)$ is the number of real parameters to specify an n-dimensional mixed state. For real amplitudes $f(n)=n(n+1)/2$, for complex amplitudes $f(n)=n^2$, and for quaternionic $f(n) = n(2n-1)$. For propagation, etc., need $f(n1n2)=f(n1)f(n2)$, which only works for complex amplitudes.

11.3 Emergent Time
A variety of efforts have been made to find a definition of time that is somehow implicit to the main QFT and GR formalisms, whether it be a choice of vacuum for QFT in curved spacetime (and even if the spacetime is not curved [90]) which is indirectly a choice of time. Or seeking an internal time-reference in a full-GR quantum minisuperspace analysis of dust-shell collapse [74]. Or in seeking a notion of time in full general relativistic (GR) models, in the equilibrium sense, with an assumption of euclideanizability [72,73]. For the latter, the self-consistent stable solutions that were indicated showed the general utility of the euclideanizability hypothesis on emanation/propagation solutions in general (that is especially relevant, or interpretable, when the system is in equilibrium). In none of these efforts, however, was there success in identifying some internal notion of time, time, it seems, is an added construction, and this is consistent with the results shown here, where we find that time is likely an emergent construct.

11.4 Emergent Evolution and Emergent Universal Learning
We see that the definition of the emanator process is not known, but that consistency arguments (such as achirality constructed from a collection of chiral emanations) lead to a certain set of forms. And that consistency with the $\{\alpha, \pi\ C_\infty\}$ relation imposes further constraints on the form of the emanator. What is hypothesized is that the emanator is selected for maximal information transmission, thus emergent itself under that criterion. Let's now consider the maximal information transmission idea from the receiving end, e.g., maximal information receiving, or learning, in this context. If we turn to the information geometry analysis of learning in neural nets [91-93] (which uses differential geometry) we obtain a

fundamental origin for statistical entropy (Shannon Entropy), and we identify optimal learning processes, based on expectation/maximization, that involves two-steps (as the name suggests) that may be done according to two fundamentally different conventions, e.g. the optimal learning involves four types of step, or is doubly chiral, consistent with the emanator 4-chiral processes described [94]. The potential applications of these results to Trigintaduonion encoded neuromanifolds is beyond the scope of this paper.

11.5 Objective Reduction, Zero-Divisors, and possible origins of Planck 's constant

A new mechanism for objective reduction [95,96] is also indicated by the way π enters the theory as a maximum anti-phase amount comprising part of the maximal perturbation propagation. Consider in the context where there is a 'classical' trigintaduonion path in a congruence of paths (a flow-line description). On the classical path in the congruences, we have α calculated using a $+\pi$ maximal anti-phase, but this could also occur with $-\pi$ maximal anti-phase as well, thus we could have a $\pm\pi$ phase toggle when a zero divisor is encountered in the 32D propagation (given the perturbations extending outside the 10D somewhat into the entire 32D). The zero-divisor discontinuity requires the field to reformulate a new 'consistency' with the 32D algebraic propagation (and 64D and higher, as well), which could have the result that since the prior π phase had the discontinuity, then it must toggle to the other, negative, phase, e.g., objective reduction may occur as a zero-divisor phase-toggle event.

11.6 Matter appears in emanator theory as a meromorphic residue of a zero-divisor

We see matter as meromorphic residue precipitation, in amounts of one quantum given by a precursor to Planck's constant h*. The meromorphic residue winding number is also notable in that it gives an integer that stays constant in the meromorphic region. This raises the possibility that elementary particle attributes might encode by way of different winding numbers, with reference to their different winding numbers at residues, but this will not be discussed further here.

11.7 What is the precise form of emanation with possible multi-card 'hands'?

So far have a deck with 72 or 78 cards when considering 'dealing a hand' of a particular size for a particular order emanation (order=number of em-steps=size of 'hand'). When we sum over the one-step emanations we have a very close match with α. When we sum over multistep emanations this will potentially be improved further. What is not apparent is what the maximal size hand is (if any) that is dealt in the emanator sum variant that considers all sizes of hands? (without replacement we theoretically have 78). The terminology of card and hand (or spread or flop) is not chosen arbitrarily since there is a precise match-up with the size of the tarot deck (78) and its decomposition via four suits (chiralities), etc. What is the maximal card flop in tarot spreads? Generally it is less than 10 or 11. Perhaps this should be taken as a hint of what will eventually be shown mathematically to be the case (for maximal information transmission). It might be that the maximal for size (hand) is dictated by $\alpha_0^{-1} + \alpha_1^{-1} + \cdots$ with upper bound, α^{-1}, given by $\{\alpha, \pi\}$ relation. In this way we know the size, composition of the deck, and how many cards are dealt in a (maximal) hand. So … not only does god play dice with the universe, but his wife plays cards…

11.8 Where's the geometry?
So far we have an explanation of standard model and h and α, but no clear explanation of G and no better understanding of gravitation. Trigintaduonion emanation shows the standard model as a direct outcome, thus a higher temperature GUT theory for unification is not required (although may still be a fundamental stage, possibly with echoes imprinted on the cosmological data and on boundaries). A theory consistent with avoidance of GUT theory and the need for inflation, is where we get inflation from conformal bounce having the same conditions. The starting point of this theory is noting that the early (Big Bang) universe and the entropy-death universe both lose all matter and thus all matter scale, and become conformally invariant. Roger Penrose has pointed this out on a number of occasions with his conformal bounce hypothesis (Conformal Cyclic Cosmology) for the early Big Bang [97-99]. Penrose makes an excellent point about the oddity of geometry not equilibrated when matter/radiation is equilibrated. This is explained in a cyclic universe when conformal death leads to conformal birth (big bang), because we can get a perfect start like inflation without inflation. Once again, however, geometry and matter/radiation, although influencing each other, appear to enter the formalism in fundamentally different ways.

With the emanation theory we have the standard model Lie algebra acting on a space (possibly creating that 'information-space' Prolog-style merely thru consistent accessing), here the algebra acting on the unit norm trigintaduonion 'base' and, over time, arrives at an 'average' of sorts of all the T_{chiral} shifts that have acted upon it, all at the 'edge of chaos', such that an emergent manifold construct appears, providing geometry (and entropy via neuromanifold constructs [94]). The role of string theory, via holographic hypothesis (ADS-CFT relation [100]), may be critical to evaluation of such complex boundary conditions, such as with black hole thermodynamics and big bounce cosmology 'boundaries'.

The geometry side of emanation theory does not result from the action of the repeated emanator product directly, but from the accumulated product in the T base that results. Geometry is, in effect, emergent (projected) on the T_{base} 'space' of the T_{em} product action. Geometry appears as a manifold construct in both space-time curvature (where it is locally given with the standard model action) and as an intrinsic entropic property, via neuromanifold 'geodesic' motion being equivalent to, and possibly the origin of, the minimization of the relative entropy (and maximization of entropy, the 2nd Law) [94]. Setting aside thermodynamic issues in this discussion, this puts the Lagrangian formulation with standard model terms, and Hilbert action for GR, into better perspective. The representation of the geometry via the Hilbert action for GR suffices with maximal extension in whatever causally connected domain of interest. So, we've got the existing QFT in CST space-time formulations in the black hole exterior, for example. We may have the resolution at the black hole horizon causal boundary via String Theory on the surface (using Ads/CFT relation and related holographic hypothesis [100,101]).

We describe repeated chiral product action on the trigintaduonion spinor space. The emanation process, consisting of a chain of chiral trigintaduonion products, leads to a Lagrangian variational formalism *with the standard model*. The origins of the parameters of the model are beginning to be understood as well. Apparently state information memory/inertia is carried via the manifold curvature response to the matter density, where 'G' is the linkage for the balance on this 'learning' process. Presumably the G learning rate is set for optimal learning, e.g., maximal information flow, and as such its value may eventually be clarified theoretically.

Chapter 12. Eucatastrophe

Maximal information propagation as an emergent construct appears to require two forms of propagation, an early hypercomplex 'emanation' that reduces to a chiral 10D propagation in a 32D trigintaduonion space, and standard propagation with complex propagators (consistent with the quantum deFinetti relation) operating inside that 10D propagation of geometry and gauge field. From the 'emanation' stage we see the maximum dimensionality and fractal limits provide the fundamental constants that then imprints upon the emergent geometry and gauge field, including giving rise to the constants α and π. The origin of α has been a long-standing mystery. So much so that the central role of α in modern physics is literally engraved in stone, the tombstones of Sommerfeld (which displays $e^2/\hbar c$, which is α) and Schwinger (which displays $\alpha/2\pi$) for example. Its origin has eluded physics for over a century, and appears to reside in the algebra of trigintaduonions.

Emanator Theory results from a hypothesized maximal information propagation and this means maximal analyticity, maximal domain, etc. As a process, Emanator theory is also hypothesized to operate up to "the edge of chaos" to permit maximal perturbation (noise) domain. When taken with the results showing that Emanator theory is Martingale, thus has well-defined limits, we then must wonder if there are well-defined multi-scale (fractal) limits. In other words, is there a relation that would tie the micro scale constant (truly) fundamental constants (as they are counted in the 22) with the cosmological scale 'constants' that have settled out, at macro scale, in the current evolution of the Universe? In this context, the Gravitational constant G is hypothesized to be a multiscale fractal coupling parameter.

In seeking a deeper theory we build on the sum-on-paths with propagator formulation to arrive at a sum-on-emanations with emanator formulation. Propagation in a complex Hilbert space, however, in a standard QM/QFT formulation, requires the propagator function to be a complex number (not real or quaternionic, etc., [19]). This prohibits what would otherwise be an obvious generalization to hypercomplex algebras. In order to achieve this generalization, we have to introduce a new layer to the theory, one with universal emanation involving hypercomplex algebras

121

(trigintaduonions) that is hypothesized to project to the familiar complex Hilbert space propagation with associated fixed elements (e.g., the emanator formalism projects out the observed constants and group structure of the standard model). The 'projection' is an induced mathematical construct, like having SU(3) on products of octonions, but here it we be the standard model U(1)xSU(2)xSU(3) on products of emanator trigintaduonions. Thus, in this book a unified variational formulation is posed, one that arrives at alpha as a natural structural element, among other things, uniquely specified by the condition of maximal information emanation.

Synopsis – Frodo Lives

Tolkien wrote of eucatastrophes [102], perhaps he anticipated the constructive role of emergent phenomena in maximum information transmission.

Image of subway graffiti saying Frodo Lives ca 1960.

References

[1] Winters-Hilt, S. Feynman-Cayley Path Integrals select Chiral Bi-Sedenions with 10-dimensional space-time propagation. Advanced Studies in Theoretical Physics, Vol. 9, 2015, no. 14, 667 – 683. dx.doi.org/10.12988/astp.2015.5881.

[2] Feigenbaum, M. J. (1976) "Universality in complex discrete dynamics", Los Alamos Theoretical Division Annual Report 1975-1976

[3] Sommerfeld, A., *Atombau und Spektrallinien* (Friedrich Vieweg und Sohn, Braunschweig, 1919).

[4] Wolfgang Pauli – Nobel Lecture. NobelPrize.org. Nobel Media AB 2021. Tue. 2 Mar 2021. <https://www.nobelprize.org/prizes/physics/1945/pauli/lecture/>

[5] Feynman, R.P. QED: The Strange Theory of Light and Matter. Princeton University Press. p. 129. (1985) ISBN 978-0-691-08388-9.

[6] Winters-Hilt, S. Unified propagator theory and a non-experimental derivation for the fine-structure constant. Advanced Studies in Theoretical Physics, Vol. 12, 2018, no. 5, 243-255. https://doi.org/10.12988/astp.2018.8626.

[7] This Is Spiñal Tap: A Rockumentary by Martin Di Bergi, 1984.

[8] Gogberashvili, M., Octonionic electrodynamics, J. Phys. A: Math. Gen. 39 7099.

[9] Chanyal, B.C., P. S. Bisht and O. P. S. Negi, Generalized Split-Octonion Electrodynamics, 2010, arXiv:1011.3922v1.

[10] Kravchenko, V., Quaternionic equation for electromagnetic fields in inhomogenous media, arXiv:math-ph/0202010.

[11] Smith, Jr., F.D. Standard Model plus Gravity from Octonion Creators and Annihilators, Quant-ph/9503009.

[12] Pushpa, P.S. Bisht, T. Li, and O. P. S. Negi, Quaternion Octonion Reformulation of Quantum Chromodynamics, Int. J. Theor. Phys., 2011, Vol 50, 2, pp 594-606.

[13] Mironov, V.L., and S. V. Mironov, Associative Space-Time Sedenions and Their Application in Relativistic Quantum Mechanics and Field Theory, Applied Mathematics, 2015, 6, 46-56.

[14] Chrisitianto, V., F. Smarandachey, A Derivation of Maxwell Equations in Quaternion Space, April, 2010 PROGRESS IN PHYSICS Volume 2.

[15] Winters-Hilt, S. The 22 letters of reality: chiral bisedenion properties for maximal information propagation. Advanced Studies in Theoretical Physics, Vol. 12, 2018, no. 7, 301-
318. https://doi.org/10.12988/astp.2018.8832.

[16] Winters-Hilt, S. Theory of Trigintaduonion Emanation and Origins of α and π. Researchgate 05/24/20.

[17] Conway, J.H. and D.A. Smith, On Quaternions and Octonions: their geometry, arithmetic, and symmetry, A K Peters, Wellesley, Massachusetts, 2005.

[18] RCHO(ST) Hypothesis, http://www.meta-logos.com/HTQG_100811.pdf , 2011.

[19] Caves, C.M., C.A., Fuchs, R. Schack. Unknown quantum states: The Quantum de Finetti Representation. J. Math. Phys. 43, 4537 (2002).

[20] Gunaydin, M. and F. Gursey. Quark structure and the octonions. J. Math. Phys., 14, 1973.

[21] Hurwitz, A. (1923), "Über die Komposition der quadratischen Formen", Math. Ann., 88 (1–2): 1–
25, doi:10.1007/bf01448439, S2CID 122147399.

[22] Winters-Hilt, S. Chiral Trigintaduonion Emanation Leads to the Standard Model of Particle Physics and to Quantum Matter. Advanced Studies in Theoretical Physics, Vol. 16, 2022, no. 3, 83-113.

[23] Dixon, G.. Division algebras: octonions, quaternions, complex numbers and the algebraic design of physics. Kluwer Academic Publishers, 1994.

[24] Furey,C.; Towards a unifed theory of ideals; arXiv:1002.1497v5 [hep-th] 25 May 2018.

[25] Harmon, P.M., Editor, Scientific Letters and Papers of James Clerk Maxwell, Vol. II, Cambridge University Press, 1995, pp. 570–576.

[26] Hankins, T.L., Sir William Rowan Hamilton, Johns Hopkins Press, 1980, pp. 316–319.

[27] Dyson, F.J., Feynman's proof of the Maxwell equations, Am. J. Phys., 1990, 58, 209-211.

[28] Dyson, F.J., Feynman at Cornell, Phys. Today, 1989, 42 (2), 32-38.

[29] Hughes, R.J., On Feynman's proof of the Maxwell equations, Am. J. Phys., 1991, v. 60(4), 301.

[30] Silagadze, Z.K., Feynman's derivation of Maxwell equations and extra dimensions. Annales de la Fondation Louis de Broglie, 2002, v. 27, no.2, 241.

[31] Manogue, C.A. and J. Schay, Finite Lorentz Transformations, Automorphisms, and Division Algebras, Hep-th/9302044.

[32] Manogue, C.A. and A. Sudbery, General solutions of covariant superstring equations of motion, Phys. Rev. 40 (1989) 4073.

[33] Cederwall, M., Octonionic particles and the S7 symmetry, J. Math. Phys. 33 (1992) 388.

[34] Lohmus, J., E. Paal, and L. Sorgsepp. Nonassociative Algebras in Physics. Hadronic Press. 1994.

[35] Moreno, G. The zero divisors of the Cayley-Dickson algebras over the real numbers. q-alg/9710013. 1997.

[36] Baez, J.C., "The Octonions," math.ra/0105155.

[37] Winters-Hilt, S. Fiat Numero: Trigintaduonion Emanation Theory and its Relation to the Fine-Structure Constant α, the Feigenbaum Constant C∞, and π. Advanced Studies in Theoretical Physics Vol. 15, 2021, no. 2, 71 - 98 HIKARI Ltd, www.m-hikari.com https://doi.org/10.12988/astp.2021.91517.

[38] McMullen, Curtis T. 2000. The Mandelbrot set is universal. In The Mandelbrot Set, Theme and Variations, ed. T. Lei, 1–18. Cambridge U.K.: Cambridge Univ. Press. Revised 2007

[39] Gilson, J. https://www.researchgate.net/publication/2187170

[40] https://www.wolframalpha.com.

[41] Fritzsch, Harald (2002). "Fundamental Constants at High Energy" . Fortschritte der Physik. 50 (5–7): 518–524. arXiv: hep-ph/0201198 . (https://arxiv.org/abs/hep-ph/ 0201198) (https://arxi v.org/abs/hep-ph/0201198).

[42] Bogoliubov, N.N.; Shirkov, D.V. (1959). The Theory of Quantized Fields. New York, NY: Interscience.

[43] Gaździcki, Marek; Gorenstein, Mark I. (2016), Rafelski, Johann (ed.), "Hagedorn's Hadron
Mass Spectrum and the Onset of Deconfinement", Melting Hadrons, Boiling Quarks – From
Hagedorn Temperature to Ultra-Relativistic Heavy-Ion Collisions at CERN, Springer
International Publishing, pp. 87–92.

[44] Richtmyer, R. D. Principles of Advanced Mathematical Physics. Springer, 1978.

[45] Collins, J. Renormalization: An Introduction to Renormalization, the Renormalization Group and the Operator-Product Expansion. ISBN-13: 9780521311779

[46] Winters-Hilt, S. Meromorphic precipitation of quantum matter with dimensionful action. May 2021. DOI:10.13140/RG.2.2.32294.24640.

[47] Briggs, K. A precise calculation of the Feigenbaum constants. Mathematics of Computation, Vol. 57, Num.195, July 1991, pages 435-439.

[48] https://en.wikipedia.org/wiki/Sedenion

[49] Erdeyli, A. Asymptotic Expansions. 1956 Dover.

[50] Erdeyli, A. Asymptotic Expansions of differential equations with turning points. Review of the Literature. Technical Report 1, Contract Nonr-220(11). Reference no. NR 043-121. Department of Mathematics, California Institute of Technology, 1953.

[51] Carrier, G.F, M. Crook and C.E. Pearson. Functions of a complex variable. 1983 Hod Books.

[52] Feynman, R.P. Space-Time Approach to Non-Relativistic Quantum Mechanics. Rev. Mod. Phys. 20, 367 – Published 1 April 1948

[53] Feynman, R.P. (1950) Mathematical Formulation of the Quantum Theory of Electromagnetic Interaction. Physical Review, 80, 440-457. https://doi.org/10.1103/PhysRev.80.440

[54] Feynman, R.P. and Hibbs, A.R. (1965) Quantum Mechanics and Path-Integral. McGraw-Hill, New York.13.

[55] G. Gabrielse, Hanneke, D., Kinoshita, T., Nio, M., & Odom, B. (2007). Erratum: New determination of the fine structure constant from the electron g value and QED (Physical Review Letters (2006) 97 (030802)). *Physical review letters*, *99*(3), [039902].

[56] Koide, Y., Nuovo Cim. A 70 (1982) 411 [Erratum-ibid. A 73 (1983) 327].

[57] Sumino, Y. (2009). "Family Gauge Symmetry as an Origin of Koide's Mass Formula and Charged Lepton Spectrum". Journal of High Energy Physics. 2009 (5): 75. arXiv:0812.2103.

[58] Winters-Hilt, S. "Machine-Learning based sequence analysis, bioinformatics & nanopore transduction detection". ISBN: 978-1-257-64525-1. (2011).

[59] Asaka, T., Shaposhnikov, M. (2005). "The νMSM, Dark Matter and Baryon Asymmetry of the Universe". *Physics Letters B*. **620**: 17–26.

[60] Karagiorgi, G.; Aguilar-Arevalo, A.; Conrad, J. M.; Shaevitz, M. H.; Whisnant, K.; Sorel, M.; Barger, V. (2007). "Leptonic CP violation studies at MiniBooNE in the (3+2) sterile neutrino oscillation hypothesis". *Physical Review D*. **75**: 013011.

[61] Cailler, C. 1917. Archs. Sci. Phys. Nat. ser. 4, 44 p. 237.

[62] Girard, P.R.. The Quaternion group and modern physics. Eur. J. Phys. 5 (1984): 25-32.

[63] Synge, J.L. Quaternions, Lorentz Transformations and the Conway-Dirac-Eddington Matrices.

126

[64] Misner, Charles W., Thorne, K. S., & Wheeler, J. A. Gravitation. Princeton University Press, 2017. ISBN: 9780691177793.
[65] Penrose, R., W. Rindler (1984) Volume 1: Two-Spinor Calculus and Relativistic Fields, Cambridge University Press, United Kingdom.
[66] Hiroomi Umezawa: Advanced Field Theory: Micro, Macro, and Thermal Physics, American Institute of Physics, 1993, ISBN 978-1563960819
[67] Collins, J. Renormalization: An Introduction to Renormalization, the Renormalization Group and the Operator-Product Expansion. ISBN-13: 9780521311779.
[68] Gibbons, G.W and M. J. Perry. Black Holes in Thermal Equilibrium. Phys. Rev. Lett. 36, 985 (1976).
[69] Hawking, S.W. Black holes and thermodynamics. Phys. Rev. D 13, 191 (1976).
[70] Hawking, S.W., Don N. Page. Thermodynamics of black holes in anti-de Sitter space.
Comm. Math. Phys. 87(4): 577-588 (1982-1983).
[71] Whiting, B.F.. Black holes and gravitational thermodynamics. Class. Quantum Grav. 7 15 (1990).

[72] Louko, J., S. Winters-Hilt. Hamiltonian thermodynamics of the Reissner-Nordström-anti-de Sitter black hole. Phys Rev D Part Fields 54(4):2647-2663. doi: 10.1103/physrevd.54.2647 (1996).
[73] Louko, J., J. Z. Simon, and S. Winters-Hilt. Hamiltonian thermodynamics of a Lovelock black hole. Phys Rev. D Vol. 55, Num. 6 (1997).
[74] Friedman, J.L., J. Louko, and S. Winters-Hilt. Reduced phase space formalism for spherically symmetric geometry with a massive dust shell. Phys Rev. D Vol. 56, Num 12 (1997).
[75] DIRAC, P.A. M., The cosmological constants, Nature, 139, pp. 323, 1937.
[76] Winters-Hilt, S. Emanator Theory using split octonions is Manifestly Lorentz Invariant and reveals why the fundamental constant $\hbar$ should be so small. Advanced Studies in Theoretical Physics, 2023.
[77] Karlin, S. and H.M. Taylor. A First Course in Stochastic Processes. 2nd Ed. Academic Press. 1975.
[78] Weyl, H., To the theory of gravity, Annalen der Physik, 359, pp. 117-145, 1917.
[79] Eddington, A., Preliminary note on the masses of the electron, the proton, and the Universe, Proc. Cambridge Philos. Soc.,27, pp. 15-19, 1931.

[80] Narlikar, J.V. The structure of the universe, Oxford Univ. Press, 1977.

[81] Cetto, A. L. De La Pena, E. Santos, Dirac's large-number hypothesis revised, Astron. Astrophys., 164, pp. 1-5, 1986.

[82] RAYCHAUDHURI, A.K., Theoretical cosmology, Clarendon Press, Oxford, 1979.

[83] JORDAN, P., The origin of stars, Der Verlag, Stuttgart, 1947.

[84] SHEMI-ZADEH, V.E. Coincidence of large numbers, exact value of cosmological parameters and their analytical representation, arXiv:gr-qc/0206084, 2002.

[85] PEACOCK, J.A., Cosmological physics, Cambridge Univ. Press, 1999.

[86] ANDREEV, A.Y., B.V. KOMBERG, Cosmological parameters and the large numbers of Eddington and Dirac, Astron. Reports, 44, pp. 139-141, 2000.

[87] RAY, S., U. MUKHOPADHYAY, P.P. GHOSH, Large number hypothesis: A review, arXiv: 0705.1836, 2007.

[88] VALEV, D. Evidence of Dirac Large Number Hypothesis. Proceedings of the Romanian Academy, Series A, Volume 20, Number 4/2019, pp. 361–368.

[89] Shawcross, G. Aperiodic Tiling. https://grahamshawcross.com/2012/10/12/aperiodic-tiling/.

[90] Winters-Hilt S, I. H. Redmount, and L. Parker, "Physical distinction among alternative vacuum states in flat spacetime geometries," *Phys. Rev. D* 60, 124017 (1999).

[91] Amari S; Dualistic Geometry of the Manifold of Higher-Order Neurons. Neural Networks, Vol. 4(4), 1991:443-451.

[92] Amari S: Information Geometry of the EM and em Algorithms for Neural Networks. Neural Networks, Vol. 8(9), 1995:1379-1408.

[93] Amari S and Nagaoka H: *Methods of Information Geometry.* 2000. Translations of Mathematical Monographs Vol. 191.

[94] Winters-Hilt, S. Informatics and Machine Learning. Wiley Publishing. 9781119716747, Sept. 2021.

[95] Diósi, L. (1989). "Models for universal reduction of macroscopic quantum fluctuations". Physical Review A. **40** (3): 1165–1174.

[96] Penrose, Roger (1996). "On Gravity's role in Quantum State Reduction". General Relativity and Gravitation. **28** (5): 581–600.

[97] Penrose, R. (2006). "Before the Big Bang: An Outrageous New Perspective and its Implications for Particle Physics" (PDF). Proceedings of the EPAC 2006, Edinburgh, Scotland: 2759–2762. https://accelconf.web.cern.ch/e06/PAPERS/THESPA01.PDF.

[98] Gurzadyan, V. G. and Penrose, R (2013). "On CCC-predicted concentric low-variance circles in the CMB sky". Eur. Phys. J. Plus. 128 (2): 22.

[99] Gurzadyan, V. G. and Penrose, R. (2018). "Apparent evidence for Hawking points in the CMB Sky". arXiv:1808.01740.

[100] Maldacena, J.. The Large N limit of superconformal field theories and supergravity.
Advances in Theoretical and Mathematical Physics. 2 (4): 231–252.

[101] Susskind, L. "The World as a Hologram". Journal of Mathematical Physics. 36 (11): 6377–6396. arXiv:hep-th/9409089.

[102] Tolkien, J.R.R. (1990). *The Monsters and the Critics and Other Essays*. London: HarperCollinsPublishers.

[103] Winters-Hilt, S. Classical Mechanics and Chaos. (Physics Series: "Physics from Maximal Information Emanation" Book 1.)

[104] Winters-Hilt, S. The Dynamics of Fields, Fluids, and Gauges. (Physics Series: "Physics from Maximal Information Emanation" Book 2.)

[105] Winters-Hilt, S. The Dynamics of Manifolds. (Physics Series: "Physics from Maximal Information Emanation" Book 3.)

[106] Winters-Hilt, S. Quantum Mechanics, Path Integrals, and Algebraic Reality. (Physics Series: "Physics from Maximal Information Emanation" Book 4.)

[107] Winters-Hilt, S. Quantum Field Theory and the Standard Model. (Physics Series: "Physics from Maximal Information Emanation" Book 5.)

[108] Winters-Hilt, S. Thermal & Statistical Mechanics, and Black Hole Thermodynamics. (Physics Series: "Physics from Maximal Information Emanation" Book 6.)

[109] Winters-Hilt, S. Emanation, Emergence, and Eucatastrophe. (Physics Series: "Physics from Maximal Information Emanation" Book 7.)

[110] Jackson, J.D. Classical Electrodynamics, 2nd Edition. Wiley 1975.

[111] Lorentz, Hendrik Antoon (1899), "Simplified Theory of Electrical and Optical Phenomena in Moving Systems" , *Proceedings of the Royal Netherlands Academy of Arts and Sciences*, 1: 427–442

[112] D'Alembert, Jean Le Rond (1743). Traité de dynamique.

[113] Laplace, P S (1774), "Mémoires de Mathématique et de Physique, Tome Sixième" [Memoir on the probability of causes of events.], Statistical Science, 1 (3): 366–367, JSTOR 2245476.

[114] Winters-Hilt S. Topics in Quantum Gravity and Quantum field Theory in Curved Spacetime. UWM PhD Dissertation, 1997.

[115] Winters-Hilt S. Emanator Theory is shown to be an optimal Martingale process at the fractal edge of chaos, where the Gravitational constant G is hypothesized to be a multiscale fractal coupling parameter. Advanced Studies in Theoretical Physics, 2023.

[116] Landau, Lev D.; Lifshitz, Evgeny M. (1969). Mechanics. Vol. 1 (2nd ed.). Pergamon Press.

[117] Goldstein, Herbert (1980). Classical Mechanics (2nd ed.). Addison-Wesley.

[118] Fetter, A.L and J.D Walecka, Theoretical Mechanics of Particles and Continua, Dover (2003).

[119] Percival, I.C. and D. Richards. Introduction to Dynamics. (1983) Cambridge University Press.

[120] Arnold, V.I. Ordinary Differential Equations. MIT Press. (1978).

[121] Arnold, Vladimir I. (1989). Mathematical Methods of Classical Mechanics (2nd ed.). New York: Springer.

[122] Woodhouse, N.M.J. Introduction to Analytical Dynamics. Springer, 2nd Edition. 2009.

[123] Bender, C.M. and S.A. Orszag. Advanced Mathematical Methods for Scientists and Engineers: Asymptotic Methods and Perturbation Theory. Springer. 1999.

[124] Robert L. Devaney. An Introduction to Chaotic Dynamical Systems. Addison -Wesley.

[125] Landau, Lev D.; Lifshitz, Evgeny M. (1971). The Classical Theory of Fields. Vol. 2 (3rd ed.). Pergamon Press.

[126] Penrose, Roger (1965), "Gravitational collapse and space-time singularities", Phys. Rev. Lett., 14 (3): 57.

[127] Hawking, Stephen & Ellis, G. F. R. (1973). The Large Scale Structure of Space-Time. Cambridge: Cambridge University Press.

[128] Peebles, P. J. E. (1980). Large-Scale Structure of the Universe. Princeton University Press.

[129] B. Abi et al. Measurement of the Positive Muon Anomalous Magnetic Moment to 0.46 ppm
Phys. Rev. Lett. 126, 141801 (2021).

[130] Einstein, A. "On a heuristic point of view concerning the production and transformation of light" (Ann. Phys., Lpz 17 132-148)

[131] Balmer, J. J. (1885). "Notiz über die Spectrallinien des Wasserstoffs" [Note on the spectral lines of hydrogen]. Annalen der Physik und Chemie. 3rd series (in German). 25: 80–87.

[132] Werner Heisenberg (1925). "Über quantentheoretische Umdeutung kinematischer und mechanischer Beziehungen". Zeitschrift für Physik (in

German). 33 (1): 879–893. ("Quantum theoretical re-interpretation of kinematic and mechanical relations")

[133] Schrödinger, E. (1926). "An Undulatory Theory of the Mechanics of Atoms and Molecules" (PDF). Physical Review. 28 (6): 1049–1070. Bibcode:1926PhRv...28.1049S. doi:10.1103/PhysRev.28.1049.

[134] Max Born; J. Robert Oppenheimer (1927). "Zur Quantentheorie der Molekeln" [On the Quantum Theory of Molecules]. Annalen der Physik (in German). 389 (20): 457–484.

[135] Dirac, P. A. M. (1928). "The Quantum Theory of the Electron" (PDF). Proceedings of the Royal Society A: Mathematical, Physical and Engineering Sciences. 117 (778): 610–624.

[136] Dirac, Paul Adrien Maurice (1930). The Principles of Quantum Mechanics. Oxford: Clarendon Press.

[137] Dirac, Paul A. M. (1933). "The Lagrangian in Quantum Mechanics" (PDF). Physikalische Zeitschrift der Sowjetunion. 3: 64–72.

[138] Feynman, Richard P. (1942). The Principle of Least Action in Quantum Mechanics (PDF) (PhD). Princeton University.

[139] Van Vleck, J. H. (1928). "The correspondence principle in the statistical interpretation of quantum mechanics". Proceedings of the National Academy of Sciences of the United States of America. 14 (2): 178–188.

[140] Chaichian, M.; Demichev, A. P. (2001). "Introduction". Path Integrals in Physics Volume 1: Stochastic Process & Quantum Mechanics. Taylor & Francis. p. 1ff. ISBN 978-0-7503-0801-4.

[141] Vinokur, V. M. (2015-02-27). "Dynamic Vortex Mott Transition"

[142] Hawking, S. W. (1974-03-01). "Black hole explosions?". Nature. 248 (5443): 30–31.

[143] Birrell, N.D. and Davies, P.C.W. (1982) Quantum Fields in Curved Space. Cambridge Monographs on Mathematical Physics. Cambridge University Press, Cambridge.

[144] Witten, Edward (1998). "Anti-de Sitter space and holography". Advances in Theoretical and Mathematical Physics. 2 (2): 253–291.

Index

A

Achiral, 42, 47, 59, 61
achiral, 2–3, 7, 11–12, 39–42, 47–50, 53–55, 59–61, 64, 75, 79, 82, 89, 101, 103, 107–110, 113
achirality, 45, 53, 57, 64, 117
Action, 74, 110, 131
action, 12, 46, 51, 59–62, 65, 72, 74–76, 83, 120, 125
additivity, 8, 11, 49, 52, 55, 109–110
adjoint, 43–44
ADS, 120
AdS, 100
Ads, 120
Aguilar, 126
Aharanov, 11
Alembert, 129
Algebraic, 129
algebraic, 2, 7, 33, 59, 61–62, 76, 80, 113, 115, 117–118, 124
Algebras, 14, 124–125
algebras, 5–10, 14–15, 18, 21, 25, 32, 62, 67–68, 95–96, 99–100, 104, 121, 124–125
Algorithms, 128
allosteric, 88
allostericity, 88
alpha, 1–3, 8, 10, 14, 32, 38, 45, 79–81, 99, 122
alphabet, 86–88
alphaperturbations, 79
amino, 87–88
Amman, 113
analytic, 10, 45–46, 62, 64–66, 71–77, 83, 89–90, 96, 99, 109–111
Analytical, 130
analytical, 128
analytically, 66, 73, 89

Analyticity, 12, 60, 76
analyticity, 2, 10, 12, 40–41, 45–47, 60–62, 65–66, 72–75, 83, 85, 109, 111, 114, 116, 121
Annihilators, 123
annihilators, 14
anticommutativity, 16
antiphase, 5, 10–11, 36–38, 40–43, 49, 55–56, 113
antisymmetric, 68–70, 93
Aperiodic, 128
aperiodic, 128
Apparatus, 99, 101
apparatus, 3, 101–102
arcana, 48, 109
artefacts, 100
artifact, 40
artifacts, 80
Associative, 123
associative, 3–4, 12, 60–61, 73, 81, 100–101
associatively, 52
associativity, 3, 16, 25, 34, 101
Asympt, 82
asymptote, 80–81
asymptotes, 81
Asymptotic, 82, 126, 130
asymptotic, 2, 67, 70–71, 81–82, 99
asymptotically, 107
asymptotics, 8
atomistic, 3
Atoms, 131
automorphism, 9, 37, 104
Automorphisms, 124
AWEC, 101

B

Babylonian, 86

133

136

Emanation, 1, 5, 8–9, 13, 32–34, 44–45, 47–49, 51, 54, 57, 61–64, 66, 79, 100–101, 107–110, 113, 124–125, 129
emanation, 1–5, 7–13, 32–36, 38, 40, 42, 44–56, 59–64, 66–67, 70, 73–77, 79, 83–85, 89–91, 95, 99–101, 103, 106–109, 113–122
emanations, 8–9, 12, 32–33, 41, 47, 49–50, 53, 57, 59–60, 79, 101, 108, 116–119, 121
Emanator, 1, 10–11, 33, 40–41, 44–45, 48–49, 57, 61, 77, 90, 96–97, 99, 101, 108–110, 114, 121, 127, 130
emanator, 1–4, 7–8, 10–12, 33, 38–39, 41, 44–50, 53–54, 57, 60–62, 64–67, 72–76, 79, 82–83, 85, 89–90, 96, 99–101, 103, 111, 116–122
emanators, 1, 41, 47–49, 61, 82–83, 91, 116
embedded, 4, 8, 10, 115–116
embedding, 115
emerge, 85
Emergence, 66, 129
emergence, 7, 10, 45, 65, 67, 85, 110–111, 113–116
Emergent, 117
emergent, 6, 8–11, 28, 32, 79–82, 84–86, 88, 113–117, 120–122
emerges, 9, 41, 76
emission, 99
emit, 4
encapsulate, 115
encapsulations, 11, 115
encloses, 73
encode, 6, 118
encoded, 118
encoding, 86–88, 95
encodings, 86
encompassed, 76
Energy, 125–126
energy, 43, 64, 84, 94, 101
Engineering, 131
Engineers, 130

engraved, 121
entire, 75, 86–87, 118
entropic, 120
Entropy, 118
entropy, 118–120
enumerated, 50
enveloping, 66, 76
equilibrated, 119
equilibria, 103, 110
Equilibrium, 127
equilibrium, 63, 69, 100, 110, 117
equipartition, 42, 55
equipartitioned, 42
equipartitioning, 39, 43
equivalence, 96
equivalent, 12, 45, 60, 93, 95, 102, 120
equivalently, 2, 15–16
Erdeyli, 126
Eucatastrophe, 121, 129
eucatastrophes, 122
Euclideanizability, 40–41
euclideanizability, 117
Euclideanization, 76, 114, 116
euclideanization, 116
Euclideanized, 115
Evolution, 86–87, 117
evolution, 1, 8–9, 37, 66, 87, 103, 116, 121
evolutionary, 87
evolves, 8
exact, 2, 5, 10, 65, 67, 88, 90, 128
Expansion, 125, 127
expansion, 43, 70–71
Expansions, 126
expansions, 36, 52
Expectation, 84
expectation, 118
Exponential, 36
exponential, 10, 37–39, 42, 52, 71–72, 76
extend, 1, 25, 76, 84, 89, 100–101
extendable, 66
Extended, 84

geometry, 3, 11, 41, 99, 101–102, 106, 115, 117, 119–121, 124, 127
Gibbons, 127
Gibbs, 13
glutamate, 88
glutamine, 88
glycine, 88
God, 4
god, 119
GPU, 26
GR, 99, 116–117, 120
Gravitation, 84, 127–128
gravitation, 119
Gravitational, 121, 130
gravitational, 83, 103–106, 127
Gravity, 14, 100, 116, 123, 128–129
gravity, 13, 116, 127
Group, 125, 127
group, 9, 15, 37, 43, 60–61, 84, 94, 99, 103–106, 111, 122, 126
groups, 64, 87
GUT, 3, 119

H

h, 43, 45–47, 57, 59, 64–65, 68, 73–74, 79, 89, 95–96, 100, 104–105, 121
Hadron, 125
Hadronic, 125
hadronic, 43
Hadrons, 125
Hagedorn, 43, 125
halting, 95
Hamilton, 13, 124
Hamiltonian, 100, 127
Hawking, 99–100, 127, 129–131
Heaviside, 13
Heisenberg, 130
Hermitian, 94
heuristic, 130
Hibbs, 126
hierarchy, 100
Higgs, 84, 88
Hilbert, 6–7, 9–10, 76, 115–117, 120–122

histories, 32
Hologram, 129
holographic, 120
holography, 131
homogeneous, 13
homogenous, 93
horizon, 99, 120
How, 62, 79
HTQG, 124
Hubble, 105
Hurwitz, 124
hydrogen, 88, 105, 130
hydrophilic, 87–88
hydrophobic, 87–88
hydrophobicity, 88
Hypercomplex, 13, 36
hypercomplex, 4, 6, 13–16, 35, 37, 39, 42, 60, 76, 89, 113–115, 117, 121
hypersphere, 6
hypotenuse, 40, 43

I

Image, 122
images, 4, 114
imaginaries, 69
imaginary, 5, 10, 12, 14–15, 19, 23, 26–27, 31–32, 34–43, 46, 48–56, 60, 62, 65–67, 69–72, 75–76, 80, 99, 114, 116
Incarnate, 66
indirect, 11
indirectly, 117
individual, 37–38, 74
individually, 47, 81
induce, 41
induced, 114, 122
inducing, 106
inequality, 45
inertia, 120
inextricable, 13
infinite, 70, 113
infinitesimal, 5, 89, 91, 93
infinitesimally, 66
infinity, 71

141

pole, 66, 73
polymer, 87–88
potential, 62, 92–93, 118
potentials, 92
precipitate, 75
precipitation, 12, 60–61, 65, 118, 125
precision, 4, 11, 25, 41, 43, 48, 56, 79–80
primary, 53
primes, 96
probability, 109, 129
Process, 9, 113, 131
process, 2–3, 7–12, 14–15, 25, 32, 45, 47, 49, 53, 60–61, 63, 77, 80–82, 85, 90, 95, 99, 101–104, 106–108, 110–111, 113–115, 117, 120–121, 130
Processes, 101, 127
processes, 7–8, 10, 25, 88, 99, 103, 107, 110, 113, 118
Product, 125, 127
product, 2–3, 7, 9, 12, 15, 18, 25, 34–37, 40, 42, 50, 52, 54, 57, 59–62, 69, 73, 75, 90–92, 96, 99, 101–102, 104, 110–111, 120
products, 9, 12, 16, 34–35, 37, 40, 50, 53, 56, 59–60, 63–64, 70, 73, 79, 96, 100, 104, 120, 122
projected, 2, 120
projecting, 40
projection, 4, 6, 8, 10, 45, 67, 73, 99, 111, 122
projections, 6, 8
projective, 5, 7, 113
proline, 87
Prolog, 119
propagate, 25, 32, 80, 96
propagateable, 9
propagated, 91
Propagating, 27
propagating, 7, 10, 27, 29–30, 32, 79–83, 85, 113, 115–116
Propagation, 32, 88, 116, 121

propagation, 4–11, 16, 26–27, 29–34, 36–38, 40–42, 51, 53, 61, 65–67, 73, 79–81, 85–89, 91, 96, 99–100, 103–104, 113, 115, 117–118, 121–124
propagations, 25, 27–30, 38, 51, 66, 80–81, 86
Propagator, 8
propagator, 1–2, 5–10, 16–17, 19–21, 23–24, 32–33, 36, 38, 64, 67, 73, 75–76, 79, 96, 104, 110, 116, 121, 123
propagators, 1, 10, 16–17, 19–20, 23–24, 48, 67, 73, 80–81, 116, 121
proper, 95, 97
protein, 87–88
proton, 105, 127

Q

QCD, 5, 10, 14, 99, 113, 116
QED, 5, 10, 70, 80, 99, 113, 115–116, 123, 126
QFT, 10, 47, 70, 99, 101, 114, 116–117, 120–121
QG, 116
QM, 70, 121
quantization, 70, 75, 116
Quantized, 125
Quantum, 5, 116, 123–124, 126–129, 131
quantum, 1–3, 7, 9, 11–14, 32, 43, 60–61, 63, 65–67, 72–73, 85, 89, 99, 101–102, 110, 114–118, 121, 124–125, 128, 131
Quark, 124
quark, 43, 84
Quarks, 125
quarks, 84
Quaternion, 13, 46, 123, 126
quaternion, 6–7, 13, 16, 90, 95
Quaternionic, 13, 123
quaternionic, 5, 9, 13–14, 95, 117, 121
Quaternions, 13–14, 124, 126
quaternions, 6, 13, 67, 90, 95, 124

145

R

radiation, 99, 119
Random, 107, 109–110
random, 2, 8, 11, 35, 65, 67, 77, 103, 107–110
randomly, 25–27, 31, 48, 80
randomness, 27–28
range, 27–28, 81, 108–109
rank, 70, 93
ranks, 9, 104
ranwalk, 108
ratio, 46, 62, 75–76, 103, 105
rational, 75
ratios, 75, 103, 105–106
RAYCHAUDHURI, 128
RCHO, 6–7, 23, 124
rczc, 27–31
Real, 5, 16, 46, 82
real, 2, 4–5, 7, 9, 11, 14–21, 23–24, 26–38, 42–43, 46, 49–54, 56, 61–67, 71–73, 76–77, 79–82, 89–91, 104, 115, 117, 121, 125
real component, 31
Reality, 4, 129
reality, 7, 32, 80, 82, 106, 124
Reals, 6, 14
reason, 57, 83
receiving, 113, 117
Reduced, 127
reduced, 19, 21, 25, 28–30, 33–34
reduces, 6, 16, 18, 49, 121
Reduction, 118, 128
reduction, 25, 118, 128
reductions, 72
redux, 89
region, 72, 83, 118
regions, 53, 70
regrouping, 34
regular, 90
regularization, 10, 47, 99, 114, 116
reified, 72, 110
Relation, 125
relation, 5, 7, 11, 18, 31, 38–42, 44–45, 48–49, 53–54, 56, 60, 65–67,

69–70, 73, 79, 83–84, 90, 93–94, 96, 104–105, 113–114, 117, 119–121
relations, 5, 14, 37, 68–69, 106, 109, 115, 131
relationship, 13, 109
Relativistic, 123, 125–127
relativistic, 101, 117
Relativity, 76, 116, 128
relativity, 94
religion, 86
religious, 86–87
Rellich, 2, 7, 39, 43–45, 52, 114
Renormalization, 125, 127
renormalization, 43, 47, 49, 63, 114
renormalize, 10, 99
renormalized, 43, 108
Repeated, 27
repeated, 7–8, 11–12, 14, 16, 25, 27–28, 35, 37, 57, 59–61, 70, 75, 79, 82, 89, 101, 110, 120
Representation, 124
representation, 14, 25, 111, 120, 128
representations, 9
reside, 93, 121
residue, 12, 60, 62, 66–67, 73–75, 83, 89, 118
residues, 72, 75, 118
Richtmyer, 125
Riemann, 71
right, 1, 4–5, 7–9, 12, 17, 20, 24–26, 32–33, 37, 42, 47, 54, 59–60, 82, 85, 104, 113–114
rightmost, 81
Rindler, 127
rise, 5–6, 15, 35, 38, 48, 65, 67, 70, 74, 79, 94, 99, 113–114, 121
rotate, 83
rotates, 65, 67
rotation, 5, 12, 56, 60, 65, 67, 71–72, 76, 92, 94
rotations, 6, 94
Rydberg, 3

S

saddle, 72
scalar, 26, 92, 94
Scale, 130
scale, 38, 43, 54, 84, 103, 105, 119, 121
scaled, 41
scales, 30, 88, 103, 106
scaling, 18
scattering, 116
Schrödinger, 131
Schwinger, 121
Second, 40
second, 23, 29, 33–35, 50, 56, 65, 67, 90, 116
seconds, 67
Sedenion, 6, 9, 17, 25, 32–33, 104, 126
sedenion, 8–9, 16, 19–21, 23, 27, 33, 36–37, 48, 51, 53, 69, 80–82, 86–88, 90–91, 95–96, 104
sedenionic, 5
Sedenions, 6, 9, 14, 16, 19, 46, 70, 104, 123
sedenions, 5–7, 15–16, 21, 25, 27–30, 33, 36, 46, 57, 62, 68–70, 91, 95–96
select, 101, 123
selected, 5, 7, 27, 32, 42, 65, 87, 113, 116–117
selecting, 81
Selection, 113–114
selection, 7, 76, 101, 113–114, 117
selects, 63, 91
self, 40, 43–44, 113, 117
semiclassical, 32, 72
separate, 37, 48, 50–51, 55, 63–64, 80, 84
separated, 43, 55, 87
separately, 11, 51, 80–81, 83
separating, 43
separation, 3, 101
sequence, 6, 57, 59, 126
Set, 4, 45, 56–57, 73, 125

set, 4, 11, 33, 37, 45, 47, 56–57, 67, 69–70, 74, 80, 82, 85–86, 90, 94, 114, 117, 120, 125
sets, 75
Shannon, 118
Shaposhnikov, 126
Shawcross, 128
Shirkov, 125
simulation, 107–108
simulations, 107
simultaneously, 14
singular, 2, 62, 75, 89
singularities, 61, 75, 130
sink, 46, 62
source, 46, 62, 81
sourceless, 61
sources, 61
Space, 115–116, 123, 126, 130–131
space, 1–2, 4–11, 13, 32, 42, 63, 66, 80, 83, 85–87, 89–91, 102, 115–117, 119–123, 127, 130–131
spaces, 15, 27
Spacetime, 129
spacetime, 9, 86, 113, 115, 117, 128
spans, 105
special, 17, 20–21, 24, 94
species, 83
specific, 5, 53, 85, 90–91
Spectrum, 125–126
speed, 105
sphere, 63, 75
spherically, 127
Spin, 123
spin, 95
Spinor, 94, 127
spinor, 94, 102, 120
spinorial, 3, 94–95
Split, 15, 89, 96, 100
split, 2, 47–48, 57, 61, 66, 87–91, 96, 115, 127
split formulation, 66
Split Octonion, 123
splitting, 15, 53, 87, 115
splittings, 15
Stability, 100

147

149

www.ingramcontent.com/pod-product-compliance
Lightning Source LLC
Chambersburg PA
CBHW050459190326
41458CB00005B/1361